The Role of Pleasure to Improve Tourism Education

Alejandra Zuccoli • Maximiliano E. Korstanje

The Role of Pleasure to Improve Tourism Education

Springer

Alejandra Zuccoli
Department of Economics
University of Palermo
Buenos Aires, Argentina

Maximiliano E. Korstanje
Department of Economics
University of Palermo
Buenos Aires, Argentina

ISBN 978-3-031-21582-7 ISBN 978-3-031-21580-3 (eBook)
https://doi.org/10.1007/978-3-031-21580-3

This Springer imprint is published by the registered company Springer Nature Switzerland AG
The registered company address is: Gewerbestrasse 11, 6330 Cham, Switzerland

This new, original, and highly readable text provides a necessary stimulus to those engaged with tourism education in a post-pandemic world. The disruption of recent years is explored as are the challenges of past approaches to the teaching of tourism, with this new text offering a fresh and critical lens on how the global phenomenon of tourism is delivered in our classrooms and lecture halls.

Prof. Alan Fyall
Visit Orlando Endowed Chair of Tourism Marketing, University of Central Florida, USA

This great book provides a bold philosophical focus on Tourism education and andragogy globally, highlighting the challenges and opportunities for the tourism education stakeholders globally. Alejanda Zuccoli and Maximilliano Korstanje examine the depth of intellectual ideologies challenging traditional approaches to tourism education and examining the future of tourism education in the new normal. In his main argument, the tenets of Tourism education need to evolve with the trends in the sector so that tourism education remains relevant, impactful and also liberating. This editorial project is an important addition to knowledge and makes an important contribution to the tourism industry globally. This book will not only give a unique contribution to knowledge but will practically have great impacts on scholars and educators globally.

Dr. Shem Wambugu Maingi
Kenyatta University, Kenya

The *Role of Pleasure to Improve Tourism Education,* which is authored by Prof. Alejandra Zuccoli and Maximiliano Korstanje, is not only a book but a gift to the tourism community. Tourism education is very important as the whole industry depends on it for a sustained future as well as for students, researchers, educators, and practitioners to get the knowledge of how to teach tourism and manage and market the industry effectively. The book brilliantly addresses key issues and challenges in tourism education while opening our eyes to the challenges in managing and marketing tourism education after the pandemic. Congratulations to the authors for this great achievement!

Prof Vanessa GB Gowreesunkar
Anant National University, Ahmedabad, India

Introduction

The problem of tourism education is far from being new. From its outset, the discipline debated between two contrasting poles. On the one hand, the business and profits are inevitably entwined not only to the industry but also to global trade. On the other hand, tourism is understood as a social fact that revitalises psychological and social frustrations occurred in daily life (Andereck & Nyaupane, 2011; Korstanje & George 2021; Korstanje 2022). To the epistemological debate accompanying the evolution of tourism research, one must add the current crisis in tourism education. Such a crisis includes – but it is not limited to – the public/private funding and budgets, the lack of interest of students in tourism-related careers and professional frustrations once students earned their degrees (Cooper & Shepherd, 1997; Hsu, Xiao & Chen 2017; Cho & Kang 2006), without mentioning the divergence between teaching and researching which was certainly aggravated by the arrival of metrics and the culture of publishing or perish (Zhao 1991; Airey et al, 2015; Edelheim, 2020). Some voices have alerted on the gap generated between theory and practice (Ritchie, Burns & Palmer, 2005). To fill the gap, this book interrogates furtherly the problem of tourism education from different angles. The curricula are mainly designed following the old traditional paradigms, which have been molded based on a much deeper Cartesian dualism. Having said this, classic education prioritises cognitive performance over emotions. The present book reflects the main advances in the Joy Labs, a department of the University of Palermo, Argentina. Joy Labs have combined as a set of an experiment named as PANCOE method where students are systematically subject to pleasurable experiences and activities such as cooking, tasting cultural dishes and products, or baking pieces of bread (only to name a few). Whilst PANCOE stimulates students' academic performance through pleasurable activities, the dropout rates have slumped down as well as the academic performance has been notably improved in those participants who take part in the experiment. Although chapters forming this editorial project can be read separately, they are integrated into a common thread argumentation. The scholastic philosophy, most probably influenced by Stoicism, developed a negative connotation revolving around pleasure. Medieval philosophers adopted the stoic viewpoint relegating pleasure to a peripheral position. The pleasure was seen as a disruptive disposition that corrupts the mind. Over the recent years, the classic higher education curricula have overlooked the importance of pleasure – as well as well-being – to foster academic production (Smith & Diekmann, 2017; Uysal et al 2016). Furthermore, educators in the field of tourism toyed with the belief that rational planning detects and solves all global dangers that may very well affect the tourism industry. The quantitative-led methods were enthusiastically embraced as valid sources to gather valuable information (Baggio & Klobas, 2017; Dwyer, Gill & Seetaram, 2012). The turn of the twentieth century has brought some uncontemplated risks which included terrorism, political violence, economic downturn as well as the rise of new virus outbreaks as COVID-19 or simply natural disasters accelerated by an unparalleled ecological crisis (Tiwari, Seraphin & Chowdhary, 2021). For instance, some commentators alerted on the urgency of updating the curricula to these new times. The crisis of the tourism industry follows a deep crisis in the higher education system given not only by the popularity of tourism but also because of disordered knowledge production (Tribe 2010; Persson-Fischer & Liu 2021).

As the previous argument is given, the first introductory chapter explores the complex intersection of education and politics. The chapter looks to respond to what extent education is an ideological instrument of domination or an opportunity towards human emancipation. We review the ancient and medieval philosophers like Epicurus and Aristotle and Roman stoics like Marcus Aurelius, Seneca and Saint Augustine. Even if ancient Greek philosophers were not reluctant to pleasure, the material asymmetries of the Roman Empire led Seneca and his colleagues to develop a negative viewpoint of pleasure. This tradition has been inherited by medieval philosophy (i.e. Descartes and Hobbes). A young Descartes – plausibly moved by a strong rationalism – showed a knee-jerk reaction to the role played by pleasure in human education. Per his stance, rationality (mind) should govern and rule over the arbitrariness of emotions. His legacy is well-known by the Cartesian dualism which dominated the modern higher education curricula for decades. However, one must speculate that medieval university has nothing to do with modern establishments. The second chapter deals not only with the invention of university that was strictly constrained to be a place for the ruling elite but also with the current problems of education in the West. The Second World War end, associated with the massification of media, created serious imbalances in the higher education system. The modern university not only received other low-ranked classes and groups but also standardised the curricula towards the democratisation of education. Theories in pedagogy are in this chapter placed into the foreground. To some extent, some scholars lamented that beyond the democratisation of education lies a much deeper process of commoditisation where free citizens become simple consumers. From Emile Durkheim onwards, philosophers have objected to what they dubbed as the education in crisis which includes higher dropout rates, excessive carrot-and-stick approaches, budget cuts and students' distress and frustration. All these problems are in-depth discussed in the second chapter which concentrates efforts to describe the evolution of modern universities and the impacts of globalisation in higher education. We debate a philosophical quandary mainly based on the rivalry between public and private education. The third chapter goes into the fields of tourism education. Of course, tourism education repeats the same problems as Western higher education. To wit, the Cartesian dualism gravitated not only to the curricula formation but also to the consolidation of tourism as a maturated discipline. From its inception, tourism evolved according to two clear paradigms. On the one hand, the founder's parents stressed the importance of measuring dynamics with the end to protect the interests of involving stakeholders. To a major or lesser degree, tourism education acted as a catalyst to form the next workforce, as well as specialists who plan the future of the industry. This economic-centred tradition monopolised not only a whole portion of curricula but also what has been published in leading journals. Education serves to find and eradicate the future global dangers that threaten the industry. The figure of forecasting occupies a central position in this process. On the other hand, a critical tradition emerged to support tourism as a rite of passage, a very important institution founded in the core of Western civilisation. For these scholars, educators should struggle to introduce critical reasoning to confront the economic-based paradigm. The fourth chapter synthesises the problems and challenges of tourism education in Argentina. The crisis prompted by the COVID-19 pandemic opened the doors to image new steps in tourism education. Unfortunately, Argentina lacks of a firm background to transfer professionals the necessary learning skills to cope with dangers like this in the future. Additionally, we analyse the syllabuses content of some leading universities. The obtained results show how the syllabuses' content lacks courses dedicated to global governance, risk perception and resiliency. The fifth and sixth chapters reflect the main outcome obtained with the study case baptised as PANCOE, a successful experiment headed by Joy Labs, at the University of Palermo. In PANCOE, which consisted of 18 meetings, students are bombarded with pleasurable feedback which included cooking, tasting dishes and other cultural activities. Those students who take part in PANCOE have reduced the dropout rates whilst increasing their degrees. Based on Epicurus' legacy, PANCOE stimulates academic performance through the stimulation of pleasure and joy. Originally, PANCOE was designed to standardise the academic curriculum of students coming from neighbouring countries. PANCOE deals successfully with emotionally deprived students in which case it is a method suggested to be applied in a post-COVID-19 context. The final chapter gives some tips to improve higher tourism education in the decades to come. Having said this, PANCOE starts from the need to re-discuss the conflictive relations between students and lecturers which today is far from being resolved. In addition, these states of discrepancies accelerate the student`s frustration. This moot point invariably leads us to the final epilogue, where we dissect the conflictive relations between students and lecturers. Based on Maximiliano Korstanje's ethnographies and pedagogic fieldworks conducted in 2012 and 2013, we hold the thesis that behind students' disengagement there lies a covered sentiment of fear. Students are often pressed to earn their degrees, but the labour market offers a grim landscape. Excessive working hours, psychological distress and low-paid jobs are some of the daily issues that students face once they are recruited by tourist firms. In this epilogue, there still is a source of inspiration to reconsider new techniques oriented to enhance students' engagement in the constellations of tourism and hospitality as well other sub-service sectors.

References

Airey, D., Tribe, J., Benckendorff, P., & Xiao, H. (2015). The managerial gaze: The long tail of tourism education and research. *Journal of Travel Research*, *54*(2), 139–151.

Andereck, K. L., & Nyaupane, G. P. (2011). Exploring the nature of tourism and quality of life perceptions among residents. *Journal of Travel research*, *50*(3), 248–260.

Cho, M. H., & Kang, S. K. (2006). Past, present, and future of tourism education: The South Korean case. *Journal of Teaching in Travel & Tourism*, *5*(3), 225–250.

Cooper, C., & Shepherd, R. (1997). The relationship between tourism education and the tourism industry: Implications for tourism education. *Tourism recreation research*, *22*(1), 34–47.

Dwyer, L., Gill, A., & Seetaram, N. (Eds.). (2012). *Handbook of research methods in tourism: Quantitative and qualitative approaches*. Edward Elgar Publishing.

Edelheim, J. (2020). How should tourism education values be transformed after 2020?. *Tourism Geographies*, *22*(3), 547–554.

Hsu, C. H., Xiao, H., & Chen, N. (2017). Hospitality and tourism education research from 2005 to 2014:"Is the past a prologue to the future?". *International Journal of Contemporary Hospitality Management*, *29*(1): 141–160

Baggio, R., & Klobas, J. (2017). *Quantitative methods in tourism: A handbook*. Channel View Publications.

Korstanje, M. E. (2022). Tourism imagination: a new epistemological debate. *Current Issues in Tourism*, ahead of print, 1–13.

Korstanje, M. E., & George, B. (2022). *The Nature and Future of Tourism: A Post-COVID-19 Context*. Apple Academic Press.

Persson-Fischer, U., & Liu, S. (2021). The impact of a global crisis on areas and topics of tourism research. *Sustainability*, *13*(2), 906.

Ritchie, B. W., Burns, P. M., & Palmer, C. A. (Eds.). (2005). *Tourism research methods: integrating theory with practice*. Cabi.

Smith, M. K., & Diekmann, A. (2017). Tourism and wellbeing. *Annals of tourism research*, *66*, 1–13.

Tiwari, P., Séraphin, H., & Chowdhary, N. R. (2021). Impacts of COVID-19 on tourism education: analysis and perspectives. *Journal of Teaching in Travel & Tourism*, *21*(4), 313–338.

Tribe, J. (2010). Tribes, territories and networks in the tourism academy. *Annals of Tourism Research*, *37*(1), 7–33.

Uysal, M., Sirgy, M. J., Woo, E., & Kim, H. L. (2016). Quality of life (QOL) and well-being research in tourism. *Tourism Management*, *53*, 244–261.

Zhao, J. L. (1991). A current look at hospitality and tourism education in China's colleges and universities. *International Journal of Hospitality Management*, *10*(4), 357–367.

Contents

Chapter 1
Towards a Theory of Education

Abstract Education should be defined as a process oriented to transmitting values, knowledge and traditions to the next generation. Since the ancient days, cultures have struggled with considerable efforts and resources to better their education system. This means that education varies in culture and time. Ancient stoics have developed an uncanny sentiment of mistrust revolving around the figure of pleasure. This disposition was invariably adopted by Roman philosophers who witnessed the material asymmetries and radical contradictions of the Roman Empire. Once collapsed the Empire, the main doctrines of Stoics were bestowed by Medieval philosophers who manifested a great hostility against pleasure and wellbeing. In this way, modern education and the higher education system cemented on a Cartesian dualism which dissociated the body and the mind. Emotions and feelings were not only relegated to a peripheral position but also subordinated to Western rationality. The present book chapter discusses critically the tenets of Western education from ancient times to postmodern relativism.

Keywords Stoics · Education · West · University · Modernism · Medieval age · Pleasure

1.1 Introduction

The earlier introductory chapter gave readers a snapshot of the problems of tourism education. Complementarily, the present chapter explores the contradictions and limitations of modern rationality in the current curricula. One of the main limitations of education today seems to be not only the monopolization of the Cartesian dualism -a dominant paradigm- which offers a dissociation between rationality and emotions, but also the reluctance to accept pleasure and well-being as key factors towards a more integrated pedagogy. From Aristotle onwards and probably much earlier, philosophy has debated critically to what extent the person knows, without mentioning the ways or under what methods persons know better or more. Over the years it is important to add, many philosophers confronted with the complex

A. Zuccoli, M. E. Korstanje, *The Role of Pleasure to Improve Tourism Education*, https://doi.org/10.1007/978-3-031-21580-3_1

interplay between politics and education. To date, different voices have alerted the limitations and problems of modern education (Apple, 2012; Holmes, 2018) which encompass a lack of funds and higher costs in the private universities and schools, overcrowded institutions, vocation problems mainly associated with the lack of opportunities in the labour-market, disinterest to embrace innovative methods by professors who are often subject to low-salaries and excessive working hours or the hegemony of liberal market to set the agenda of educators as well as the inadequacy of curricula to resolve the inter-class material asymmetries generated by global capitalism (Klees, 2020; Peters, 2004). The fast growth of mass media and digital technologies, adjoined to the multiplication and dissemination of information, has accelerated a crisis in the educational system as never before (Compaine, 2001). This raises a more than pungent point: is education a disciplinary mechanism of control or a valid way towards individual emancipation?

Let us remind readers that the term *educere* was originally used in Ancient Rome to denote two contrasting meanings. At a first look, there was a positive connotation associated to "emancipation", understood as the freedom of spirit, but at the same time, the term signalled to the act of feeding cattle, which suggested a type of submission (or degradation). Most certainly, the figure of domination (or alienation) occupied a central position in Marxist and Structuralist approaches (Sharp et al., 2017) causing a serious influence in modern social sciences. As Lawton & Gordon put it, critical theory in education fields have emphasised on the urgency to decipher all given paradigms and great narratives as ideological instruments of control, and in so doing paradoxically it laid the foundations towards a radical (methodological) relativism that invariably led education into an unparalleled state of disorientation (Lawton & Gordon, 2005). Having said this, the present introductory book chapter gives readers a fresh insight into the fields of education, as well as its current barriers without mentioning the problems in modern university curricula. In the first section, the tenets of ancient philosophy as well as the ambiguous nature of pleasure is placed into the foreground. We explore the main arguments of Stoic philosophers of the calibre of Aristotle, Epicurus, Seneca, Marcus Aurelius and Saint Agustin. Here two assumptions should be made. On one hand, whilst ancient Greek philosophy kept a conservative position revolving around pleasure, Romans developed a radical critique against the liberation of low human instincts. The essence of pleasure -unless dully regulated- was demonised by Roman philosophers as the sinew of all human vices. Romans not only envisaged the injustices and inequalities of a large (and sometimes unjust) Empire but also described how hedonism corrupted the social ties in their days. In respect, Roman philosophers were central to the formation of medieval and modern philosophies of education. Part of the Cartesian dualism, which demonised not only pleasure and love but also suggested the authority of mind over the body, is scrutinised in the second section, comes from Aristotle. Rene Descartes, Leibniz and other philosophers focused their attention on the dualism between the mind and the body, the mind and the matter. Needless to say, this Cartesian dualism greased the rails of modern education as well as the curricula from its inception. The third section is reserved to debate the weakness and strengths of modern philosophy to approach the education system crisis. Radical voices such as Hannah Arendt are put in dialogue with other liberal ones as Bertrand Russell or Richard Rorty. We proffer a short review on the advances of social sciences (which include psychology, sociology, pedagogy and anthropology) to answer the open questions that are unresolved to the date. Because of time and space, we are unable to review all authoritative voices who had adopted education as its main object of scrutiny. We have done, rather, a short review of the main thesis in which case some of them have certainly criticised the unresolved intricacies and contradictions culturally enrooted into Cartesian dualism. Lastly, a short review of the constructivism in education gives an open insight to the next chapters, which we -the authors- hope enjoying the readers.

1.2 Ancient Philosophy and Education

In the ancient world, philosophy has devoted considerable efforts to discussing the nature of education and its impact on the community. For ancient Greek philosophers, the city and politics were inextricably intertwined. Knowledge –like arts and other activities- is neither revealed in a pure form nor can be grasped by humans. Plato's cavern seems to be a brilliant example that explains how the excess of knowledge intoxicates the human soul. For Plato, the cavern exhibits a perfect metaphor to show how the human soul works. In addition, the essence of truth remains always inexpugnable to the human eye, unprepared to reach total knowledge (Deleuze & Krauss, 1983). In a similar line of inquiry, Aristotle's treatise *On Education* deals critically with the intersection of reasoning which helps to grasp the surrounding world and how habits mould what the man knows. As he puts it, repetition cultivates the human spirit into good habits, but what is more important, humans often pay exaggerated attention to the figure of *excess* as a disruptive behaviour which gradually invites them to immoderation (hubris) (Aristotle, 1953; Robb, 1943). Aristotle's legacy seems to be still timely nowadays illuminating the ways for many philosophers in the threshold of time (Arthur, 2019; Curren, 2000).

As the previous backdrop, Aristotle –most certainly like Socrates- alerts on the negative role played by an excess of pleasure on the learning process. It is important to note the inter-link of pleasure and learning was vital in Aristotelian texts. Although he recognizes the good desire motivates men to learn something new and in so doing the world turns more pleasurable, no less true is that unlimited pleasure should be understood as a supervening end in itself. The learning process is also defined as an act of conscience that acquires knowledge from the external world (ethos). As a pure end that empties once it is achieved, pleasure -lest closely monitored- drives gradually towards hedonism (Munteanu, 2004; Woerther, 2008).

It is worthy of mention that not all ancient philosophers agreed with Aristotle and his stoicism. Epicurus of Samos was an authoritative voice who introduced the figure of pleasure (joy) assisting people to get away from fear and pain. In Epicurus, the purpose of philosophy is mainly oriented to help people to reach happiness. He strongly believed that the importance to live a self-sufficient life should be accompanied by the presence of good friends. One of the bases of our pain is mainly associated with death denial. He was a staunch enemy of the Platonic idea of reasoning. Furthermore, this argument places Epicurus conceptually pitted against Aristotle –though they never exchanged correspondence or bolstered a dialogue-. He ultimately affirmed the paradox whether of education is determined by personal experience rather than abstract reasoning. Not only pleasure is the lack of suffering, as he adheres, but it contributes positively to self-achievement, which is successfully orchestrated by the spirit, potentiates learning. In Epicurus' viewpoint, pain is mainly marked by a much deeper sentiment of fear of death. However, no less true is that death represents one of the ends of life and an unquestionable truth (Epicurus & Gerson, 1994). Over the years, Epicurus was misunderstood as a rampant exegete of a hedonist pleasure that corrupts the human soul. Bur far from that he held the thesis that happiness and freedom are only achieved through the absence of pain, but this does not mean the denial of pain. Epicurus' axiom proclaims overtly that any person should be happy only by accepting suffering as a part of life -albeit understanding the problems of the excesses and vices-. He distinguishes two types of pleasures. At a closer look the moving pleasure takes room when a person successfully satisfies his desires at a first instance, but this type of pleasure is tied only to an ongoing movement because once satisfied it returns. What is equally important, static pleasure contemplates living the blessedness of life whilst consecrating the human spirit to self-fulfilment (Mitsis, 2014; Rist, 1972; Warren, 2004). It is important not to lose sight of the fact that Greek philosophy energetically condemned hedonism and chaos, warning on the problems and limitations to succumb to the hegemony of hedonist (capricious) wishes. This simply happens because Ancient Greece was a civilization fraught with internal and external disputes and conflicts marked by a constant state of warfare which explained the aversion of philosophy for death, violence and war, and the empathy for the hard work. The real misconception of the legacy left by Greek philosophy (above all in his interrogation on pleasure) is certainly given by the rise of Roman philosophy and the Empire (Harte & Lane, 2013). At a closer look, Romans interpreted Greek philosophy through the lens of their epoch. Since 27 BC, Rome went through rapid economic growth aligned with a rampant military expansion through Europe and Asia. Hence, the Empire –as an overarching power- indexed some peripheral (economies) as rich colonies but at a higher cost. Rome came across the rise of internal material asymmetries which caused serious economic problems in the wealth distribution process. It is not whimsical to admit that the old stoic doctrine (philosophy), which strictly focused on hard work and agriculture labour, set the pace to the vices of large Empire mainly characterised by inter-class conflict, financial speculation, hedonist and public spectacles carefully orchestrated to dominate the lower classes as well as the hegemony of a real-estate market widely dominated by the Senate and the Roman aristocracy. To put the same in bluntly, some slaves who lived in the rich household or under the auspices of the Imperial house have more benefits than free roman citizens. Roman philosophers –echoing Greek stoicism- gradually become staunch enemies of pleasure, extravagant luxuries and opulent lifestyles. After all, they are direct witnesses of how pleasure and luxury wreaked havoc in the social scaffolding (Bloomer, 2011; Bonner, 2012). The decline of the Roman Empire is explained by the excess (on pleasure) of higher classes and the ruling elite. From different angles and moments of the Roman Empire, Seneca, Marcos Aurelius and Saint Agustin interrogate furtherly on the power of ethics and the urgency for persons to stave off laziness. Seneca (the younger) is born in the 4 B.C century. Although he did not know the glory of August Emperor, he served as advisor of Nero who blamed him to take part in a conspiracy to kill him. In consequence, Seneca was forced to kill himself. Beyond this drama, Seneca was an important stoic philosopher who had contributed substantially to the understanding of education. He was widely influenced by Epicurus though he was originally educated in the ancient stoic doctrines. His main thesis holds that education leads the man to ethical perfection moderating through reasons some destructive emotions such as pain, grief and anger (Seneca, 2011). In his philosophical model, the correct use of wealth depends not only on living an active and contemplative life but also on the capacity to practice poverty (Lawton & Gordon, 2005; Volk & Williams, 2006). Doubtless, Seneca was inspired by the economic and social injustices that occurred during the zenith of the Roman Empire. Marcus Aurelius, instead, as Emperor was pressed to live the start of the end of the great imperial authority. Marcus Aurelius was born on 26 April of 21 A. C. He is crowned emperor in 161 A. C succeeding Adrian, whilst pressed to conduct two wars against Germanic Tribes. His triumph, which was immortalised in the film Gladiator, mainly marked a serious economic downturn for Rome. As a philosopher, he devoted considerable resources and time to deciphering the earlier stoic texts. He argues convincingly that the philosophy should return to the axioms of Cato the Younger; having said this, Marcus Aurelius believed that discipline (not pleasure) occupied a central position in moral education which is the touchstone towards a peaceful life (Aurelius, 2013). As Professor Van Ackeren (2012) eloquently highlights, Marcus Aurelius should be well recognised as an emperor-philosopher who interrogated on the philosophical intrigues through the constellations of power and humility. He showed how the work of philosophy should be simple and modest, taking part in the seduction of politics and vain ostentation. Starting from the premise that the relationship between philosophy and politics is at least troublesome and complex, Marcus Aurelius overtly accepted that education should be centred on the objective arbitrations of ethics.

Unlike the other above-discussed philosophers, Saint Augustine (Augustine of Hippo) is passively contemplating the fall of the Roman Empire. He is a monk born on 13 November of 354 A. C who is witnessing the risks and problems of hedonist pleasure in politics. One of his major fears is the corruption of the body. All human beings are perfect creations of God originally dotted with the interplay between the soul and body. There is, anyway, a dramatic fight between the drives toward sensual pleasures (like sex, banquets) and the perfection of soul for any person is prepared by God. In his books, the human soul should govern rationally the pretensions of bodies whilst keeping away from sin. As a superior entity, indivisible, the soul regulates the lower instincts of man (Augustine, 2010). With the benefits of hindsight, this distinction paved the way not only for Cartesian dualism but also for the consolidation of medieval philosophy and a clear aversion to the pleasures of the flesh

(Ingham, 2005; Williams, 1970). Once finally collapsed the Roman Empire, Europe lived an epoch of darkness and extreme violence characterised by great political atomization and fracture (feudalization) where feuds struggled with each other all the time. Medieval philosophy had a great influence on modern education as well as the trivialization of pleasure to potentiate human self-achievement. For the sake of clarity, this point will be developed with more accuracy in the next section.

1.3 Medieval Philosophy and Cartesian Dualism

As stated in earlier parts of this book chapter, medieval philosophers were extremely worried about the advance of pleasure in politics and philosophy (Mills, 2006). Medieval days were mainly marked by an extreme violence as well as a manifest hostility by "the Other". The figure of pleasure and happiness were seen with strong mistrust or simply overly rejected in the philosophical circles (Lawton & Gordon, 2005). The cultivation of pleasure, far from leading the person to virtue, engenders countless vices and problems for the human soul (Inwood, 2005). As a result of this, they had the urgency to come back to the legacies of Roman philosophers who cautioned their readers on the importance of ethics. Whatever the case may be, their fear of "the pleasure of the body" or the dark human instincts, medieval thinkers paved the way for the creation and consolidation of a strange axiom which highlighted the axiom that the mind (which incarnated the perfection of the soul) should control the body (which signalled the irrationality of emotions) (Kaempfer, 2018; Shapiro, 2018). This dualism inspired a young French philosopher, Rene Descartes to punctuate that there is a body-mind dualism given by the presence of modes, which exhibit the ways how the substances co-exist. The body-mind dissociation explains the intellectual abstraction that understands (grasp) the substance beyond its connection with the body. For some reason which is very hard to precise here, modern philosophers not only centred on the figure of education but also borrowed the old conceptions and debates that originated in Roman Stoicism. In the fields of politics, the Roman Empire has fallen, but new European powers launched to colonize the world. From the Kingdom of Spain and Portugal to France and the United Kingdom, since the fifteenth century, the European nations indexed and modelled the world to their image and likeness (Korstanje & George, 2021). The surfacing chasm between emotionality and rationality left by Roman stoics legitimated not only the European colonialism claiming a so-called superiority of European rationality over other aboriginal voices but also affirmed the centre-periphery dependency (Ureña, 2017). The savages, like women, were widely portrayed in novels and travel writings as irrational agents often prey of basic drives and egoism whilst the European man was instructed by the universal law to re-educate them in the ideals of enlightenment (Pagden, 2005; Pratt, 2007; Simpson, 2007). As discussed in earlier sections, Rene Descartes was a French philosopher who was born on 31 March of 1596. His influence originally comes from Aristotelianism. Without any doubt, this led him to revive the spirit of Aristotle and ancient Stoics but to some extent, Descartes strayed far from the influence of his predecessors. Per his viewpoint, he offers more than an interesting model which centred on the division between mind and body. Cartesian dualism not only accompanied a whole portion of his theology but also permeated other philosophical models as well. In this respect, Descartes argued convincingly that the sensual (external) world was an invention of the mind. Turning back to extreme solipsism, he says that the self should disengage from the world even negating its existence. But in essence, what cannot be denied is the existence of the self. If the only valid truth is that I exist, God is the self elevated to the infinite. To put things bluntly, the conception of reality is only given at the time mind separates ontologically from the body. He coins the term Cogito, Ergo Sum (I Think therefore I am) whilst putting the ego as the centre of thinking activity. In this way, the self disengages from the material environment whilst adopting rational contemplation. It is difficult to resist the impression that Descartes' dualism which illuminated the path of modern philosophers and educators, applauded enthusiastically the belief that the body and the mind are separated entities but working together. What is more important as an entry in this debate, the body, as well as emotions, subordinates to the rational mind (Descartes, 2012). His genius and work can be compared to other philosophers such as Gottfried Leibniz, John Locke, or even Thomas Hobbes. In the next lines, we shall approach Hobbes, Leibniz and of course John Locke's contributions.

As a leading philosopher, Hobbes, in fact, notably influenced by the British civil war and the political violence that traversed England, tellingly agrees with Descartes that education centres on a long-lasting peace. The Leviathan (state), from where emanates the legal monopoly of force, regulates civil life (avoiding the War of all against all) but in some instances, conflict emerges. Education assists the state in the creation of a moral spirit mainly based on western rationality and science. Hobbes revives the long-dormant debates left by Descartes whilst alerting that human irrationality mines the pillars of politics. Like Roman stoics, the good man should not embrace pure pleasure since it feeds greed and envy. If I want what my neighbour possesses, also it is not strange that my neighbour will crave my possessions. This philosophical dilemma ushers civility in an inflexible gridlock. To solve this, Hobbes introduces the figure of Leviathan which embodies the political authority of the state and the law. But was is more important, we cannot imagine this Leviathan's sovereignty far from what seems to be legitimated by moral education (Hobbes, 2016, 2019). Most certainly, Hobbes' detractors blame his philosophy to be a bit authoritarian but far from this, he offered a brilliant diagnosis of his days. As Teresa Bejan (2010) notes, Hobbes not only accelerated an academic revolution in medieval philosophy but also exerted a caustic criticism on the educational system of his epoch. Per his stance, Hobbes lamented the degree of corruption in universities as well as the dominant academic paradigms. As a matter of supreme urgency, educators should start with a moral and educational reform to re-educate lay citizens. Otherwise, the arbitraries and injustices derived from politics –far from disappearing- place the political authority in jeopardy. With the benefits of perspective, Hobbes was a critical voice who punctuated the deficit and limitations of university curricula. In sum, civic education is vital to frustrate the return of a despotic king to the throne. Here we can build a bridge between Hobbes (and of course Descartes) and Leibniz. In his readings, Leibniz, who knows well Hobbes' thesis, supported the rationalism of Descartes on education. Since education is ruled by the principle of contradiction and reason, which means that all things are God, no less true is that the human experience makes from the universal mysteries and objects of knowledge. However, as he clarifies, this experience is not stable in time and very well it is subject to change and contradiction. To wit, the figure of monads, which is paradoxically formed by mind and matter, contains the same dualism expressed by Descartes (Leibniz, 1989).

As the previous argument is given, John Locke never centred his object of study in education or pleasure in the strict sense of the word but he discussed both – though tangentially- in depth. At a first glimpse, he defines the self as a conscious thinking thing (entity) finely dotted of spirituality and substance. This self, one must confess, appears to be sensitive to pain and pleasure as well as happiness or sadness. As a staunch rival of Cartesian Dualism, he toys with the belief that the learning process is main marked by the experience, (not the rationality or abstraction). What any man knows results from different propositions which are internalised through the individual experience. Any idea contains reflections which derive directly from rationality and sensations. All men are empty cabinets who are filled according to their experiences and biographies. Nobody can understand what cannot be previously experienced (Locke, 1887). A-la-Kant, Locke strongly believes that education moulds gradually the man making useful or not his previous experiences (Moseley, 2014; Tuckness, 2010). Last but not least, these two academic positions not only are contrasting and opposite but also fiercely fight in the fields of education even to the date. Both, Descartes and Locke laid the rational foundations towards a new understanding of education and its impacts on society.

1.4 Modern Philosophy and Education

Not surprisingly, it is impossible to start a section in modern philosophy and education without introducing readers to Bertrand Russell. Unlike other philosophers, Russell emphasised on the influence of logic on language. As a good philosopher of mathematics, he becomes the founding parent of analytic philosophy (in the twentieth century). Comparable to Ludwig Wittgenstein or Gottlob Frege, Russell put idealism under the critical lens of scrutiny. Originally influenced by Leibnitz, Russell reminds us that the advance of science is ensured by the will to doubt, which means the possibility to interrogate –test and negate- the dominant discourses. The sense of discovery is formulated by the possibility to test and discard previous hypotheses. The man of science accepts the idea any given knowledge is not only temporal but also an object of discussion and scrutiny. In this respect, education should be understood as an active attempt to liberate the man from the submission of parental and religious authorities. Only love, as well as the quest for truth, is enough to transform society whilst struggling to dethrone the authoritarian cultural mainstream values. Far from domesticating the citizen, education should serve to emancipate the free spirit inventing a system free from political repression. Per Russell, education has been commoditised (if not ruined) the capitalist values of exchange and consumption. Hence we witness how there is a philosophical dilemma between the individual and the citizen, as he adheres. The question of whether education trains good individuals or citizens is of paramount importance in Russell's argumentation. The modern education system was originally designed to fabricate useful docile citizens but not good citizens who care for the community. To solve the problem, educators and policymakers should acknowledge that students are mirrored by emotions, by love and pleasure (joy) which activates the mere acts of knowing. His main thesis holds that in the learning process, power, emotions and knowledge are inextricably intertwined (Russell, 2009, 2013). He goes on to say: *"the issue between citizenship and individuality is important in education, in politics, in ethics, and metaphysics. In education, it has a comparatively simple practical aspect, which can be to some degree considered part from the theoretical issue"* (Russell, 2013: 12). In a nutshell, one of Russell's main contributions lies in the fact that the perfection of man is not triggered by a classic high education, but by the harmonic integration between the individual and citizen (Park, 2013). This argument penetrates in a young German philosopher Hannah Arendt who in 1958 published a trailblazing book which entitled "The Human Condition". In this work, she brings some critical reflections on what she dubbed as *vita activa* and *vita contemplativa* (Lat.) as two main forces framing the human condition. Whilst the former signals to the vital condition which moves us to achieve our goals the latter refers to metaphysical reflection. Having said this, man dwells very well in a world coexisting with the alterity (which means with Others). It is noteworthy that politics mediates not only among human beings but also institutions. In opposition to Hobbes, she holds the polemic thesis that violence shorts run to regulate citizens. Instead, the authority of the state situates in direct contradiction with human freedom. Hence, the power should be understood not as a submission act but as the capacity to transform the environment. In Arendt, the sentiment of reciprocity plays a vital role in impeding the Hobbesian dilemma of *the War of all against all* nor as the unique condition for durable peace, this does not suffice. Modern capitalism creates an alienatory climate of domination which undermines critical thinking and subsequently the individual autonomy. The success of the state consists in introducing moral relativism as the touchstone of society. In hedonist societies, pleasure maximization takes the lead whereas negating pain as a bad emotion. Invariably this leads citizens to idolatry and a state of ongoing insecurity (Arendt, 2013). In consonance with this, the education system faces an unparallel crisis simply because students are widely trained (not educated) as potential workforce for the liberal market. The nature of education conflicts with the productive means of capitalist society, as Arendt concludes (Biesta, 2010; Gordon, 2018). This raises a question difficult to answer: to what extent good education does impact democratic life? As Gert Biesta (2010) brilliantly observes, one of the main Arendt's contributions to the politics of education rests on the axiom that alienated citizens do not contribute positively to society. The triumph of democracy in Western culture is invariably entwined to a properly-educated citizenry simply because education should be seen as the preparation of citizens for future political participation in daily life.

Ultimately, a section fully dedicated to education and philosophy cannot be closed without reviewing American philosopher Richard Rorty. In close dialogue with Dewey and liberals, Rorty outlines how the importance to educate the citizen about politics. Educated in the University of Chicago, a young Rorty embraces Dewey's axioms and what experts dubbed as philosophical pragmatism. The external nature and matter have no meaning beyond the power of language. What any person knows is always engulfed in the language. We live life and know based on self-construed representations which are extrapolations of our inner world. He coins the term "mirror of nature" to explain the nature of knowledge. Rorty is convinced that the modern problems of western epistemology rest on the problems of mind to imagine the external reality. In the human mind, all beliefs are finally justified by other profound beliefs. The science and learning process is given by the self-evident foundations that help humans create beliefs whilst adapting them successfully to the world. Academic disciplines, in this direction, move in quest of partial truths which only tell a part of the entire story. This conceptual stance situates Rorty in the opposition to Kuhn. Having said this, education evokes an old paradox claimed by liberal pragmatists, whilst the lay-citizen is liberated from ignorance he is systematically constrained by the authority of what he learns (Rorty, 1990, 1999). The type of relativism introduced by American pragmatism was a serious challenge posed on modern social sciences. Once again as their predecessors, social scientists would try to respond to the following vexed question: is education an instrument of indoctrination or an opportunity or pathways towards a more just society?

1.5 Social Sciences and Education

From its inception, social sciences were motivated to find practical solutions to overcome those pitfalls that led the educative system to an inevitable crisis. Although the founding parents of social sciences (i.e. Marx, Weber, Pareto and Durkheim) have shown particular interests in education, there was no other Scholar like Emile Durkheim who offered an all-encompassing diagnosis of education crisis (Ottaway, 1955; Prus, 2011). Globally recognised and cited as one of the fathers of sociology, Emile Durkheim laid his finger on the advance of anomie as well as the disintegration of social ties as one of his more prominent objects of study. In essence, he questioned what is the premise that kept society integrated. In his different works, Durkheim called attention to the crisis the education system was facing in France. The advance of modernity, as he claims, undermines the social ties as well as the social reciprocity, accelerating a moral decomposition. Social maladies such as addictions, alcoholism, crime and alienation are part of the problems (pathologies) of the modern lifestyle. At the same time, capitalism expands social reciprocity recedes. Of course, the education crisis explains by the decline of the classic labour market. Education trains citizens to be introduced in the labour market. In the nineteenth century, Durkheim witnessed an inevitable radical transformation of capitalism and the economic market in Europe. He defines education as a means by which each society trains the future workforce (above all adult generations). Lastly, Durkheim holds that sociology should provide reliable information to resolve the crisis of education (Durkheim, 1956, 1977, 2012). Instead, philosophy, social sciences (which include psychology, sociology, anthropology only to name a few disciplines) have developed different conceptions revolving around education. From Pavlov to Piaget, for example, psychology has fleshed out its project to deal with education; a position which was not shared by anthropology or ethnology that applied educational theories in aboriginal tribes and organizations. The popes of anthropology like Bronislaw Malinowski, Margaret Mead even Evans Pritchard alluded to the importance to borrow native methods to improve the current problems in the Western educational system (Levinson & Pollock, 2016). Scholars of the calibre of Benjamin Lee Whorf (2012) described how the learning process is mainly marked by the sense of constructed reality. The cosmology is given by the set of beliefs, emotions, stereotypes we have embraced to understand the surrounding space. No matter the culture, anyone can learn something that cannot be spoken or expressed from the language. Whorf acknowledges that aboriginals who are unfamiliar with the chronological time fail to pass successfully the western tests. This notion leads Whorf and his colleagues to question the essence of human intelligence which is framed by Westernised paradigms. Rather, sociology echoed the earlier Durkheimian texts to denounce the decline of morality and education in the West. By its end, psychology centred efforts in offering an all-encompassing diagnosis for better understanding the individual process of learning as well as the institutional aims of modern education (Thorndike, 1910). In consonance with this it is noteworthy that Jean Piaget was a pioneering voice in pedagogy who delved into the fascinating fields of cognitive development. As long as the 70s and 80s decade, Piaget began with a series of experiments that are based on a child-centred approach. As he eloquently notes, children follow the same patterns and developmental models as adults to learn something. Having said this, learning is an ongoing process that resulted not only from interaction but also from individual experience. Each learner creates a type of balance between what is taught and his previous experience. However, the curricula are shaped by a set of different content which is demanded and designed to meet the needs of the labour market. Knowledge should be esteemed as a type of compensation individually negotiated but culturally conditioned (Piaget, 1966, 1975). Over the years, other scholars have alerted on education as an alienatory mechanism of control operating in two directions. On one hand, it dominates some signifying codes and ideologies which are drawn to subordinate all classes to the ruling elite (Middleton, 1963; Williamson & Cullingford, 1997). On another, education keeps the material asymmetries created by the economic system (Beach, 1999). This is the case of Basil Bernstein who in 1973 published the trailblazing book *Class, Codes and Control.* As a purely linguistic, Bernstein contends that classes –as social construes- are formed and based conforming specific linguistic codes. Whilst the higher class develops an elaborated code which allows abstract reasoning, the lower classes are strictly limited to work only with a restricted code.

The elaborated code denotes an open correlation between means and ends giving the speaker the possibility of selecting among a selective range of alternative courses of action to resolve a problem. Contrariwise, the restricted code seems to be severely limited to ritualist forms of communication. For the sake of clarity, he puts the example of a mother telling her children a story, finally, she asks him "what do you think happened?" here, a response is given by a set of alternatives emanated from child rationalization. The restricted code is verbally communicated and understood in the forms of mandates limiting the children not to imagine the future or the end of the story. Bernstein suggests that these two codes are successfully managed by higher-classes whilst the lower ones only circumscribe to the restricted code. This invariably means that the inter-class subordination is previously determined by the economic structure which permeates the language. Education legitimates –in Bernstein's analysis- through language a type of class hierarchy that is very hard to change. In consequence, classes adopt different linguistic processes that are far from being equal. In a nutshell, the lower classes are educated to passively abide (if not internalize) the dominant discourses resigning their critical thinking; Bernstein innovates in a linguistic model that describes why students coming from blue-workers homes have few degrees than other students. His thesis is that education, far from resolving, affirms the inter-class inequalities engendered by modern capitalism (Bernstein, 1973).

For the sake of clarity, it is important to mention that the 70s decade represents a turning point in pedagogy and the philosophy of education. Politically speaking the dismantling of European Empires, just after the WWII end, associated to an academic climate of criticism against the dominant classes cemented a new revolution in education in Europe and the US. Naturally, a disrupting point of divergence was originally posed by Michelle Foucault in France. To uncover Foucault's legacy, we must recognize he is not only a philosopher but a militant for the human rights of minorities. In this way, he questions critically not only the status quo but the nature of science and history as well as any other dominant narratives of his epoch. As he notes, knowledgeable associates to power to be employed to silence, marginalize or even eradicate "other marginal groups". His theory of power and knowledge has influenced notable scholars contributing directly to disciplines like psychiatry, anthropology and psychology. The production of knowledge, which only tells a part of the real story, legitimates the economic status quo through the exercise of disciplinary power which means the use of bodies` skills effectively to make citizens more obedient entities oriented to work and consumption. In using their skills, the dominant class thwarts the possibilities to head a real revolt (or revolution). Here is where education goes. As a result of this, education (if not the disciplinary power) is successfully articulated to transform citizens into objects which are manipulated with techniques that lead to rational self-control. In so doing, this power should be individually and collectively internalised and accepted. For Foucault, this occurs not only in schools or universities but also in factories, hospitals or prisons where subjects are re-socialised to renounce their real interests whilst rebuilding unilateral and closed patterns of behaviour. Ultimately, this discipline is cemented through three main mechanisms: observations, exams and a normalised judgment. The observation, always hierarchical, refers to the possibility to visualize the controlled body. This control is justified, desired and internalised when the body embraces what he dubbed as the "normalization process" which leads to normalised judgment. Finally, exams have a double purpose. They are drawn to give a certification of the learned skills whilst allowing gathering information that speaks of the observed body (Foucault, 2008, 2012, 2018, 2021). Some decades later, similar findings are obtained by French sociologists Pierre Bourdieu and Jean Claude Passeron who in 1990 published *Reproduction in Education.* In this work, they develop the concept of social reproduction as finely ingrained in the educative process. Education plays a leading role not only in reproducing the dominant narratives but also in the cultural values of the elite to keep holding the power. It is important to mention that this submission is primarily not given by the economic factor, as classic Marxists say, but with a symbolic matrix lay citizens engage. Hence, society exerts symbolic violence which exhibits the ability to ignore the social hierarchy accepting it as normal. Also, education helps recycle social inequalities conditioning one life chances through the construction of what they named as "social capital". In this vein, the legitimacy of higher classes is given by the success and velocity of reproduction education simply offers (Bourdieu & Passeron, 1990). The Cartesian dualism, as well as the ideals of Stoicism, is gradually contested by many theorists and scholars worldwide.

Theories on education have recently been multiplicated globally in social sciences but they could be very well classified in three main families: (a) structuralist school (b), rational school (c) social reproduction school. Again, let us give readers a short introduction to the commonalities and substantial differences between these schools. At a first glimpse, the *structuralist school* (or family) acknowledges that education serves to keep society well-integrated. Society is equated to a human body, mainly marked by the interaction of organs (institutions) and processes. Citizens are educated to perform not only a role but a function which is of paramount significance for the economic system. This role is finally filled in interaction with others who follows their interests (Morozumi, 2018). The education process socializes future citizens providing them with a robust framework to deal with problems, frustrations and anxieties in the labour market. Although not limited to, the structuralist school is based on quantitative-related methods (Halsey et al., 1997; Kogan et al., 2011; Segal, 2013). Of course, since each actor has its interests and aims, *the rational choice school* focuses on the role played by education to control inter-group conflicts. Scholars who take part in this academic circle holds that students (like persons) have goals that are fulfilled maximizing gains whilst minimizing costs. For that, policymakers should think of innovative curricula that psychologically motivate students. Centred on qualitative research, this school emphasizes on the carrot-and-stick approach (Bridge & Wilson, 2015; Evans, 2008; Stocké, 2019). These two academic traditions neither discuss critically the social inequalities culturally enrooted into the system nor analyze the curricula formation. To fill the gap, the social reproduction school greases the rails of the complex mechanism of education laying the foundations towards a new understanding of class ideology. To put things in other terms, scholars argue that modern education reproduces the mainstream values, narratives and stereotypes proper of higher-classes. The hegemony is cemented giving lower-classes the false hope they are taking part in something important. Beyond the techniques of the motivation in Schools and universities, several statistics and reports show two contrasting tendencies. The higher school drop-out rates apply to lower-classes as well as blue-collar workers' homes, whilst rich citizens invest in more higher education which includes Bachelors, Masters or Business MBAs and doctorates. At the same time, these authors alert that there are concrete and strong barriers for low-class members to access higher education. Far from emancipating lay citizens, education serves as an instrument of domination to keep lower-classes under control (Marginson, 2010; Olssen, 2003; Sadovnik, 1991).

1.6 Constructivism in Perspective

Recently, pedagogy and education moved through the lens of Enlightenment which means the respect for laws and absolute knowledge. The dismantling of European Empires just after the 50s decade, not only open the doors to a new epoch but also began with an epoch baptised as "deconstructivism" (Hoteit, 2015; Hoy, 1990). This postmodern movement focused its attention on the fragmentation of a unique reality, the immanent reality emanated from all universal values. In this token, deconstructivism appealed to a type of deconstruction where the dominant (European) paradigms were at least put in the lens of scrutiny. In the philosophy of education, deconstructivism shed light on important voices alerting on the urgency to give further priority to the emotional world of the subject and overlooking the established curricula (Haywood & Ghaill, 1997). This means that students –far from being homogenised agents- elaborate their psychological engagement whilst cementing a particular learning process according to their biographies as they have been culturally and individually negotiated. This movement (re-baptized as constructivism) in Education has culturally engulfed into the paradigm of relativism making a substantial and caustic critique to current educators and their forms of evaluation. Echoing the main contributions of Dewey, or Derrida, these studies look for a heterogeneous dynamic of modern education concerning individual values (Hyslop-Margison & Strobel, 2007; Larochelle et al., 1998; Sutinen, 2008). For constructivism, it is important to analyse (and of course understand) how knowledge is produced and disseminated. Far from being an objective entity, knowledge is a social construal created and protected by a privileged group. In this respect, constructivism has widely operated in all levels of education, even in higher education. The rules and protocols of each discipline determine the ways of accessibility each student has. Constructivism has devoted efforts and resources to explain the academic failure in students whilst emphasising the need of adopting non-hierarchical learning processes. For this cultural movement, the learning process is based on an individual construction of knowledge leaving behind the quest for the universal truth (Petraglia, 1998). To put things more simply, the reality of the subject constructed and not vice-versa. Reality does not exist beyond the subjective cosmology of agents. Hence, the previous cognitive construes which are proper for students play a vital role in the way they finally learn. Each student should understand what he is learning (Osborne, 1996; Splitter, 2009). The constructivist theory has been widely criticised because of its impossibility to forge formal reasoning which is paradoxically based on discipline and standardised –if not mechanical-forms of knowledge. After reading a paper, or newspapers, some students showed serious problems to make analysis content. This suggests that beyond their differences, students keep some commonalities as rational agents, constructivism simply undermines (Cole, 1992; Richardson, 2003). The philosophical quandary lies in the fact that education debates in two contrasting poles, which have not shortened the chasm. The constructivist paradigm, which is pitted against the classic curricula, abrogates for interacting learning between lecturer and student. The reflective dialogue allows a less hierarchal form of education but shorts fall in the long-turn cognitive retention. Rather, classic education defends a hierarchical and passive education with strong content curricula. The mechanical repetition of knowledge gravitates positively on the improvement of students' academic performance. Both movements, needless to say, conflict even to date (Cole, 1992; Fosnot, 2013).

1.7 Conclusion

As discussed in several parts of the present introductory chapter, the term education, which comes from (Lat) *Educere*, connotes ambiguously not only a type of intellectual emancipation but also a sense of domination and social reproduction. From its inception, the modern education system was mainly influenced by what philosophers named *Cartesian dualism* which means a clear dissociation between the body and emotions and the mind. It is safe to say that this dissociation has marked the agenda of educators over years. The chapter synthesizes how Stoic philosophy – above all in Roman Empire- develops an uncanny sentiment of mistrust revolving around the figure pleasure and joy which is finally crystallised in medieval philosophy. What is more important to debate, education is facing today a serious (and for some voices irreversible) crisis leading social sciences and pedagogy to a gridlock. We have reviewed the weakness and strengths of different scholars as well as subdisciplines trying to answer the future of modern education in an ever-changing world. Last but not least, because of the limits in space and time, we are unable to review all theorists, writers or exponents who have focused on education in their careers. To resolve this, the selection was done according to the line of a paradigm or the school each author represents instead of author by author. Having clarified those theories which approached education since ancient time to date, the real problems and contradictions of modern education today, a topic that constitutes an important part of this book, are discussed in depth in the next chapter.

References

Apple, M. W. (2012). *Can education change society?* Routledge.
Arendt, H. (2013). *The human condition*. University of Chicago press.
Aristotle, J. A. K. (1953). *The ethics of Aristotle the Nicomachean ethics*. Allen & Unwin.
Arthur, J. (2019). *The formation of character in education: From Aristotle to the 21st century*. Routledge.
Augustine, S. (2010). *The City of God, Books I–VII (The Fathers of the Church, Volume 8)* (Vol. 8). CUA Press.
Aurelius, M. (2013). *Marcus Aurelius: Meditations, books 1–6*. Oxford University Press.
Beach, D. (1999). Alienation and fetish in science education. *Scandinavian Journal of Educational Research, 43*(2), 157–172.
Bejan, T. M. (2010). Teaching the leviathan: Thomas Hobbes on education. *Oxford Review of Education, 36*(5), 607–626.
Bernstein, B. (1973). *Class, codes and control: Applied studies towards a sociology of language*. Routledge.
Biesta, G. (2010). How to exist politically and learn from it: Hannah Arendt and the problem of democratic education. *Teachers College Record, 112*(2), 556–575.
Bloomer, W. M. (2011). *The school of Rome: Latin studies and the origins of liberal education*. University of California Press.
Bonner, S. (2012). *Education in ancient Rome: From the elder Cato to the younger Pliny*. Routledge.
Bourdieu, P., & Passeron, J. C. (1990). *Reproduction in education, society and culture* (Vol. 4). Sage.
Bridge, G., & Wilson, D. (2015). Towards an interactive sociological rational choice approach to theorising class dimensions of school choice. *Policy & Politics, 43*(4), 493–507.
Cole, P. (1992). Constructivism revisited: A search for common ground. *Educational Technology, 32*(2), 27–34.
Compaine, B. M. (Ed.). (2001). *The digital divide: Facing a crisis or creating a myth?* Cambridge.
Curren, R. R. (2000). *Aristotle on the necessity of public education*. Rowman & Littlefield Publishers.
Deleuze, G., & Krauss, R. (1983). Plato and the simulacrum. *October, 27*, 45–56.
Descartes, R. (2012). *Meditation on first philosophy*. Simon and Schuster.
Durkheim, E. (1956). *Education and sociology*. Simon and Schuster.
Durkheim, E. (1977). On education and society. *Power and Ideology in Education*, 92–105.
Durkheim, E. (2012). *Moral education*. Courier Corporation.
Epicurus, L. P., & Gerson, B. (1994). *The Epicurus reader selected writings and Testimonia*. Hackett Publishing.
Evans, A. (2008). Madrasah education: Necessity or rational choice? *Harvard International Review, 30*(3), 58.
Fosnot, C. T. (2013). *Constructivism: Theory, perspectives, and practice*. Teachers College Press.
Foucault, M. (2008). Panopticism "from" discipline & punish: The birth of the prison. *Race/Ethnicity: Multidisciplinary Global Contexts, 2*(1), 1–12.
Foucault, M. (2012). *Discipline and punish: The birth of the prison*. Vintage.
Foucault, M. (2018). Discipline. In *Rethinking the subject* (pp. 60–69). Routledge.
Foucault, M. (2021). History of systems of thought. In *Language, counter-memory, practice* (pp. 199–204). Cornell University Press.
Gordon, M. (2018). *Hannah Arendt and education: Renewing our common world*. Routledge.
Halsey, A. H., Lauder, H., Brown, P., & Wells, A. S. (1997). *Education: Culture, economy and society*. Oxford University Press.
Harte, V., & Lane, M. (Eds.). (2013). *Politeia in Greek and Roman philosophy*. Cambridge.
Haywood, C., & Ghaill, M. M. A. (1997). Materialism and deconstructivism: Education and the epistemology of identity. *Cambridge Journal of Education, 27*(2), 261–272.

Hobbes, T. (2016). *Thomas Hobbes: Leviathan (Longman library of primary sources in philosophy)*. Routledge.
Hobbes, T. (2019). *The elements of law: Natural and politic*. Routledge.
Holmes, B. (2018). *Comparative education: Some considerations of method*. Routledge.
Hoteit, A. (2015). Deconstructivism: Translation from philosophy to architecture. *Canadian Social Science, 11*(7), 117–129.
Hoy, T. (1990). The moral ontology of Charles Taylor: Contra deconstructivism. *Philosophy & Social Criticism, 16*(3), 207–225.
Hyslop-Margison, E. J., & Strobel, J. (2007). Constructivism and education: Misunderstandings and pedagogical implications. *The Teacher Educator, 43*(1), 72–86.
Ingham, R. (2005). 'We didn't cover that at school': Education against pleasure or education for pleasure? *Sex Education, 5*(4), 375–388.
Inwood, B. (2005). *Reading Seneca: Stoic philosophy at Rome*. Oxford University Press.
Kaempfer, L. (2018). Pleasure in the middle ages, edited by Cohen-Hanegbi, Naama, and Piroska Nagy. *Emotions: History, Culture, Society, 2*(2), 340–341.
Klees, S. J. (2020). Beyond neoliberalism: Reflections on capitalism and education. *Policy Futures in Education, 18*(1), 9–29.
Kogan, I., Noelke, C., & Gebel, M. (Eds.). (2011). *Making the transition: Education and labor market entry in Central and Eastern Europe*. Stanford University Press.
Korstanje, M. E., & George, B. (2021). The colonial voyages. In M. Korstanje & B. George (Eds.), *Mobility and globalization in the aftermath of COVID-19* (pp. 25–46). Palgrave Macmillan.
Larochelle, M., Bednarz, N., Garrison, J., & Garrison, J. W. (Eds.). (1998). *Constructivism and education*. Cambridge.
Lawton, D., & Gordon, P. (Eds.). (2005). *A history of Western educational ideas*. London.
Leibniz, G. W. (1989). The monadology. In L. Loemker (Ed.), *Philosophical papers and letters* (pp. 643–653). Springer.
Levinson, B. A., & Pollock, M. (Eds.). (2016). *A companion to the anthropology of education*. Wiley.
Locke, J. (1887). *Some thoughts concerning education*. CUP Archive.
Marginson, S. (2010). Higher education in the global knowledge economy. *Procedia-Social and Behavioral Sciences, 2*(5), 6962–6980.
Middleton, R. (1963). Alienation, race, and education. *American Sociological Review, 28*(6), 973–977.
Mills, R. (2006). *Suspended animation: Pain, pleasure and punishment in medieval culture*. Reaktion Books.
Mitsis, P. (2014). *Epicurus' ethical theory*. Cornell University Press.
Morozumi, T. (2018). Reconstruction of participation: Some implications to social studies education from a comparative study on youth participation of Sweden and Japan. *The Journal of Social Studies Education in Asia, 7*, 77–95.
Moseley, A. (2014). *John Locke*. Bloomsbury Publishing.
Munteanu, D. L. (2004). *Ancient spectator of tragedy: Facets of emotion, pleasure, and learning*. University of Cincinnati. Ph D. Dissertation Thesis. Available at https://etd.ohiolink.edu/apexprod/rws_etd/send_file/send?accession=ucin1100892095&disposition=inline
Olssen, M. (2003). Structuralism, post-structuralism, neo-liberalism: Assessing Foucault's legacy. *Journal of Education Policy, 18*(2), 189–202.
Osborne, J. F. (1996). Beyond constructivism. *Science Education, 80*(1), 53–82.
Ottaway, A. K. C. (1955). The educational sociology of Emile Durkheim. *British Journal of Sociology*, 213–227.
Pagden, A. (2005). Imperialism, liberalism & the quest for perpetual peace. *Daedalus, 134*(2), 46–57.
Park, J. (2013). *Bertrand Russell on education*. Routledge.
Peters, M. (2004). Lyotard, Marxism and education: The problem of knowledge capitalism. In J. Marshall (Ed.), *Poststructuralism, philosophy, pedagogy* (pp. 43–56). Springer.

Petraglia, J. (1998). *Reality by design: The rhetoric and technology of authenticity in education*. Routledge.
Piaget, J. (1966). The psychology of intelligence and education. *Childhood Education, 42*(9), 528–528.
Piaget, J. (1975). Comments on mathematical education. *Contemporary Education, 47*(1), 5.-15.
Pratt, M. L. (2007). *Imperial eyes: Travel writing and transculturation*. Routledge.
Prus, R. (2011). Examining community life "in the making": Emile Durkheim's moral education. *The American Sociologist, 42*(1), 56–111.
Richardson, V. (2003). Constructivist pedagogy. *Teachers College Record, 105*(9), 1623–1640.
Rist, J. M. (1972). *Epicurus: An introduction*. CUP Archive.
Robb, F. C. (1943). Aristotle and education. *Peabody Journal of Education, 20*(4), 202–213.
Rorty, R. (1990). The Dangers of Over-Philosophication--Reply to Arcilla and Nicholson. *Educational Theory, 40*(1), 41–44.
Rorty, R. (1999). Education as socialization and as individualization. *Philosophy and Social Hope*, 114–126.
Russell, B. (2009). *On education*. Routledge.
Russell, B. (2013). *Education and the social order*. Routledge.
Sadovnik, A. R. (1991). Basil Bernstein's theory of pedagogic practice: A structuralist approach. *Sociology of Education*, 48–63.
Segal, C. (2013). Misbehavior, education, and labor market outcomes. *Journal of the European Economic Association, 11*(4), 743–779.
Seneca, A. L. (2011). *Moral letters to Lucilius*. Michael Hussey.
Shapiro, L. (Ed.). (2018). *Pleasure: A history*. Oxford.
Sharp, R., Green, A., & Lewis, J. (2017). *Education and social control: A study in progressive primary education*. Routledge.
Simpson, M. K. (2007). From savage to citizen: Education, colonialism and idiocy. *British Journal of Sociology of Education, 28*(5), 561–574.
Splitter, L. J. (2009). Authenticity and constructivism in education. *Studies in Philosophy and Education, 28*(2), 135–151.
Stocké, V. (2019). The rational choice paradigm in the sociology of education. In *Research handbook on the sociology of education*. Cheltenham Edward Elgar Publishing.
Sutinen, A. (2008). Constructivism and education: Education as an interpretative transformational process. *Studies in Philosophy and Education, 27*(1), 1–14.
Thorndike, E. L. (1910). The contribution of psychology to education. *Journal of Educational Psychology, 1*(1), 5.
Tuckness, A. (2010). Locke on education and the rights of parents. *Oxford Review of Education, 36*(5), 627–638.
Ureña, C. (2017). Loving from below: Of (De) colonial love and other demons. *Hypatia, 32*(1), 86–102.
Van Ackeren, M. (Ed.). (2012). *A companion to Marcus Aurelius* (Vol. 96). Wiley.
Volk, K., & Williams, G. D. (Eds.). (2006). *Seeing Seneca whole: Perspectives on philosophy, poetry and politics*. London.
Warren, J. (2004). *Facing death: Epicurus and his critics*. Clarendon Press.
Whorf, B. L. (2012). *Language, thought, and reality: Selected writings of Benjamin lee Whorf*. MIT press.
Williams, C. (1970). St. Augustine on nature, sex and marriage by John J. Hugo. *The Thomist: A Speculative Quarterly Review, 34*(2), 330–332.
Williamson, I., & Cullingford, C. (1997). The uses and misuses of 'alienation'in the social sciences and education. *British Journal of Educational Studies, 45*(3), 263–275.
Woerther, F. (2008). Music and the education of the soul in Plato and Aristotle: Homoeopathy and the formation of character. *The Classical Quarterly, 58*(1), 89–103.

Chapter 2
Problems of Education in the West

Abstract The specialised literature punctuates that the invention of Western University was coincident with the importance to educate a privileged elite who would govern and rule the future of the next generation wisely and with moderate character. Based on the ancient meritocracy, the medieval university was a place of reflection and critical reasoning only for a few men. The expansion of capitalism, as well as the urgency to transform workers into consumers, have paved the way for the rise of the modern university while expanding the produced knowledge to other marginalised classes. As a result of this, some unknown pathologies and problems surfaced. This chapter explores the problems and future of Education in the West whilst alerting us to the challenges in the next years. The expansion of modern education has ushered students into a state of crisis which includes drug addiction, violence, higher dropout rates as well as low grades and students` lack of motivation -only to name few of them-.

Keywords Western education · Engagement · Public university · Education crisis · Lack of motivation

2.1 Introduction

In the earlier introductory chapter, we ignited a hot debate revolving around the role played by Cartesian dualism, as well as its aversion to pleasure, in the formation of education and curricula. In so doing, we have reviewed some leading authors who have historically shed light on the fields of modern pedagogy and philosophy of education. This chapter, complementarily, seems to be reserved to describe what are the practical problems and obstacles the educative system is going through in the West. These problems vary in culture and country. To set a clear example, whilst the UK has faced serious imbalances with Brexit and over-seas students (Mayhew, 2017; Tomlinson, 2019) in the US policymakers devoted efforts to reduce the rate of school and university shootings (Wike & Fraser, 2009; Borum et al., 2010). Based on the K12 School Shooting database, experts estimate that most cases of shootings

A. Zuccoli, M. E. Korstanje, *The Role of Pleasure to Improve Tourism Education*, https://doi.org/10.1007/978-3-031-21580-3_2

concentrate in those states forming the East Coast whilst the state of Nebraska, North and South Dakota report lower levels of violence (K-12 Shooting Database, 2021). Other significant issues linked mainly to the lack of funding in state schools and universities (Piché, 2015; Gillespie et al., 2001), higher student dropout rates (Tinto, 1975; Müller & Schneider, 2013), the decline of lower-class students for higher education, as well racial issues at classrooms (Mickelson, 2003) without mentioning the serious vocational crisis in students who once graduated face labour problems linked to low wages, accompanied with excessive working hours and the indifference of flexible unions to the workers´ demands (Sierpinska et al., 2008). Methodologically speaking, the present book chapter does not discuss the extensive literature formed by studies and publications in all levels of education; rather, it concentrates efforts in approaching the solutions on higher education issues. Henceforth, we begin in the first section describing the nature and evolution of modern universities in the Western hemisphere which includes Europe, the US and Latin America. For Miller and Sabapathy (2011) the question of education cannot be dissociated from ethics and responsibility combined with the efficiency in the different levels of the structure. Students should engage with the urgency of cultivating significant and global knowledge to improve society whereas society should compromise to perpetuate knowledge for next generations. At the same time, both education and society should remain open to new changes and mutual feedback. To put the same in other terms, the coin term *Open University for* describing the levels of tolerance and empathy for lay-persons the academic staff should develop. Sometimes, academicians work at an ivory tower believing that their knowledge should not be shared with lay citizens. The opposite is equally true, society seems to be closed –if not reluctant- to academic education simply because it is considered complex and futile. The role of modern university is still placed under the critical lens of scrutiny by specialists even to date. Like medieval philosophy as well as Cartesian dualism, higher education has some problems to accept wellbeing and pleasure as dominant forces towards a new more sustainable pedagogy. The goals of this chapter are double-fold. On one hand, we analyse the grass-roots levels of higher education and the present problems they face as well as the reasons operating behind the higher education crisis. On another, we ignite a hot debate revolving around the effects of COVID-19 and social distancing aggravating the crisis or giving the opportunity to a new re-foundation of the modern university and applied-research.

2.2 Education and Leisure Society

It is important to mention that from its inception economy and social science have been captivated by the study of leisure. Since Veblen and his historical sweep of industrial society to Joseph Dumanzedier, a dearth of scholars has paid attention to leisure as their main object of study (Veblen, 2017; Stebbins, 2016; Rojek, 2013). Like pleasure, leisure has been frightened by classic thinkers (Sadlek, 2014). As a closer look, leisure invariably results from the free time generated by the advance of industrialism as well as the adoption of new technological inventions. Of course, new forms of production and consumption needs a new education (Rojek, 1993). Elias and Dunning (1986) alert that the advance of capitalism has engulfed new habits, traditions and social behaviour more oriented to leisure practices. These actions are naturally embedded in new rules and emerged forms of emotionality. The public spheres have been strictly limited to impersonal relations whereas emotions can be manifested in privacy or at home. As a result of this, a more pathological manifestation of violence and looting surfaces whenever emotions cannot be dully regulated by the modern bureaucracy. For Elias and Dunning, leisure is not good nor bad, but an ambiguous instrument (like pleasure) which encourages or discourages social cohesion. For instance, Anthony Veal (2018) coins the term *leisure society* to refer to a new stage of capitalism where technological breakthrough has created an excess of free time which is absorbed by leisure practices. Echoing Dumanzedier's legacy, he argues convincingly that leisure may be well led to destructive practices unless people are educated to move into the new leisure landscapes. Per his stance, leisure should be channelled through edifying and self-developmental ways helping the agency to achieve successfully their goals. Otherwise, we run the risk to create a society where the working time without palpable sacrifice moulds selfish citizens. To some extent, leisure and economic production are inextricably intertwined. Whilst post-Fordist society abandoned the scale of production, new forms of leisure as well as education have been introduced. This invariably represents a point leisure studies fell short. Here lies a paradox, following Veal, the same technology used to have further free time can be employed to control leisure practices. In those societies -in the half-century- where there was a manifest growth in people's lifetime more money and consumption have been spent in that time. Without any doubt, this is the reason why leisure has attracted the attention of countless theorists and specialists over the recent decades. Under some conditions based on mutual regulation, leisure should not be considered as a mechanism of alienation but as a pathway towards well-being. Whilst work gives stability to citizens, leisure plays a different (but not for this less important role), it sanitises the socio-psychological frustrations that happened in daily life. Leisure, in this way, acts as an articulator of the social bond regulating inter-group and class conflict. These bad connotations revolving around leisure and pleasure have varied in culture and time (Zuzanek, 2020). Of course, the same applies to education. Lawton and Gordon (2005) hotly debate in their book [A history of Western educational ideas) the evolution and maturation of education in the Western hemisphere. Far from being a historical study, their book shows an erudite review of the ideas and the mutations of theories on education since Aristotle to date. As they eloquently note, education changes according to not only the times but also the type of society. Whereas preliterate society needs fewer rules and regulations to train its members, the industrial society imposes unilaterally a set of strict bureaucracies to educate the next workforce. This particularly may be oppressive and discouraging for students. Nevertheless, modernism has introduced a new type of methodological relativism that has directly affected modern education. Authors draw a conceptual continuum between closed and open society, albeit they do not use both terms. Some societies are restricted to bolster little dialogue or trade with foreigners. These human groups do not encourage contact with strangers whilst regulating strictly leisure practice. Lawton & Gordon put the example of Sparta. In opposition, other communities remain open to arts and foreigners encouraging travels, leisure activities and expeditions abroad. Athens seems to be a clear example of this kind of society. The question of education is equally important. In this respect, open societies impose decentralised forms of education homogenised for all citizens whereas closed one reserves education (like leisure) only to a privileged class. This conceptual model not only oscillated from one side to another in the threshold of time but also shed light on the curricula throughout ancient history to date. In this vein, historians and the historical method are of vital importance to understand education and its current crisis. To some extent, leisure and education were inextricably intertwined. In sum, the problem of education is associated with the problem of rationality or at least how the concept is culturally construed. Like education, rationality includes -of course, all rational men- but excludes irrational ones. Such an idea has changed over time and culture. Per Lawton & Gordon, education should be understood as a pedagogic process aimed at transmitting specific cultural values, traditions and knowledge to the next generation. With the rise of democratic institutions in the West, education expanded to other relegated classes, probably generating a serious crisis in the curricula. Having said this, the democratic optimism -as a leading paradigm- was cemented by what authors named "economic nationalism" which dominated education since WWII end. For scholars taking part in this paradigm, education training the future workforce to be recruited in a liberal market gives -in this way- equal opportunities for everyone. This concept was plausibly linked to the notion of continuous growth which suddenly stopped just after the 70s decade. As the authors remark, the breakdown of economic nationalism paved the way for the rise of new relative methods which confronted the ideals of Enlightenment. For example, the critical theory -based on Marxist studies- exerted a radical criticism on rationalism, as well as pluralism understanding that both concepts certainly derive from the capitalist society. What really happens, education articulates great narratives to accommodate the material contradictions among classes, public goals, the law and of course the market. At first glimpse, any market should be free when it can be regulated. Under this premise, education is not an exception. Decades later, postmodernism as a new global phenomenon not only introduced radical shifts in higher education but also global (and standardised) forms of consumption that ultimately commoditised education. This means that postmodern education escapes to the control of nation-states inculcating new rationality in future generations.

2.3 The Western University: A Short Companion

A historical description of the University in the West leads us to medieval universities in Europe (de Ridder-Symoens & Rüegg, 2003). The medieval university was the founding archetype of education in Europe. Nevertheless, the medieval university was limited to a few literate elites whilst transpiring in a clear meritocratic logic (Wilbrink, 1997). The transformation of medieval institutions was given by a radical shift in their mission and vision. In the modern universities, which are originally moulded by humanism, the structure around ideals of democratization and massification chrysalises into the enrolment of low and middle classes. Furthermore, the modern university not only homogenises mechanised-driven roles and specializations for the lay citizens but also supports the values of the modern nation-state. In sum, if the medieval university is mainly marked by the need of illuminating only a few minds whereas, the modern one embraces enthusiastically the idea of inquiry and scientific research occupies a central place to expand the produced knowledge to the entire society (Scott, 2006). Whereas modern education expands, it paradoxically becomes a well-exchangeable commodity (Doti, 2004; Tilak, 2008). As Meyer and Schofer (2007) put it, higher education, as least as it was imagined in Europe, has been rapidly exported to the world at the end of the 60s decade –once the nation-state project was ultimately consolidated-. In this respect, education accompanied the European colonial expansion in former centuries, but just after the mid of 20s century, the process was successfully consolidated. As a worldwide phenomenon, in the threshold of time higher education demand has increased incorporating new actors and classes which were historically relegated from the wealth distribution process. The classic university was a place reserved only for a small portion of population. In 1900 only a half of million students were enrolled globally whilst in 2000 this cipher rose to hundred million students enrolled in universities. This demand for higher education is particularly more present in developed nations in which case, the financial asymmetries between the global North and South are certainly enlarged. Quite aside from this, mass education has shown that just after the mid-twentieth century, the labour market demanded further skilled jobs and specializations. The expansion of the capitalist system not only generated more specialised jobs and occupations but also needed further specialization in higher education. It is noteworthy that the costs for higher education were notably reduced according to a new global world. This global pattern of expansion invariably coincides with the standardization of curricula and higher education syllabuses (Benavot et al., 1991; Kamens et al., 1996). Some studies focused on the importance to examine the expansion of the capitalist system and mass education as the origins when the current crisis takes place (Christopherson et al., 2014). The failure of mass education to make a more efficient workforce is based on what Mandler dubbed as the crisis of meritocracy (Mandler, 2020). Other scholars, rather, suggest that higher education created functional benefits for the economy that reasonably led governments to push for an aggressive standardization. Higher education permeates in other relegated classes according to the needs of the market and nation-state to engage with new consumers, not vice-versa (Witmer, 1970; Hadden & London, 1997). Here two more interesting points surface: is the Western University in Crisis? If so, why or what are the factors that accelerated such a downturn?

Christopherson et al. (2014) overtly recognize that each region or country shows its particular discrepancies and conflicts in higher education. Whilst in the US higher education is monopolised by the private sector, in some Latin American countries the central administration invests considerable resources to support public education. In the US, higher education seems to be a personal investment to what extent all costs are absorbed by students` families. This suggests that countries keep some conceptual debates beyond the fact, so to speak, whether higher education is or not a collective good. The situation in Europe appears to be more complex. Over the recent years, some universities have adopted an increase in tuition fees. What is more controversial, the higher demand for education has resulted in higher levels of young unemployment and uncertain (unstable) conditions in the labour market. To here, an inevitable question surfaces: to what extent universities educate future citizens or professionals mainly influenced by managerial logic? In this context, no less true seems to be that the concept of the civic university successfully came to stay, as the authors conclude. The opposite is equally true, state-sponsored universities are facing an unparalleled crisis simply because of serious fiscal imbalances. In this vein, Phillip Moriarty calls attention to the role of scientists (not educators) to disseminate knowledge production through different levels. As he eloquently notes, scientists should devote efforts to elucidating fundamental questions that explain how nature behaves. This means a scientist is neither an engineer nor an entrepreneur. Scientists should not be attracted to the possibility their research makes a profitable product and of course, their motivation should be centred on bettering the current understanding of things. The point here what is a valuable object of study for scientists. Moriarty exerts a caustic critique on the current Higher education Funding Council in the UK (known as HEFCE) as well as the Research Council (RCUK). Both traditional institutions are being gradually emptied and re-designed to protect the interests of private stakeholders and the market. This means that unfortunately nothing is investigated that lacks certain value for the market, as he ascertains. At the same time, professional researchers are mainly emboldened by their impacts in leading journals and published metrics instead of cultivating objectivity. The concept of science –as well as education- as a public good has set the pace for business-driven interests.

For McKay and Rowlingson (2011), it is clear how the capitalist system has generated new inequalities among classes calling attention to the incorporation of new reform in education. Beyond what it is named as the rhetoric of fairness, there lays a policy carefully drawn to reinforce already given inequalities. At a first glimpse, the success of elitism to regulate higher education consisted in imposing specific-based ideals explaining what is reserved for a few hands is more efficient than standardised and expanded institutions. This is the symbolic core of academic elitism. For McKay & Rowlingston the symbolic touchstone of elitism rests on the axiom that the state should be removed from the regulation of the private market. The evidence suggests that efforts to captivate lower-income students failed for many reasons. One of them is the fear of indebtedness lower-income families conserves. A second and most important cause reminds us how despite universities widening the quota of students (incorporating other classes) the margin of graduates remains at the same levels. Students should be supported with other qualitative assistance other than the economic incentive. Human capital, echoing Bourdieu which is replicated through the inter-class interaction, plays a key role in the education of next professionals, researchers and professors. In accordance with a similar belief, Diane Reay (2012) overtly says that universities (at least in the UK) act as spaces of reproduction of social inequality and segregation. The Browne Report, which introduces new recommendations for the future fee and tuition policies for undergraduate and post graduate students, means the triumph of money over creativity and mind. In fact, as she brilliantly holds, The Browne Report -widely applied in British universities- rests on the fact that education does not evolve without capital. Given this point, the diversity of higher education should be organized according to socio-economic and racial terms. Universities are classified in three tiers: elite university (Russell group), red bricks universities and post 1990 universities including polytechnics. Whilst the latter group is diverse in ethnicity and socio-economic background, the Russell group takes the lead preserving the same customs and classes than 40 years back. The access to higher education for middle-class students or other minority ethnicities appears to be ensured only for red bricks and post 90 universities. In this perspective, supporters of private education energetically neglect the claim that higher education in the UK is based on an elitist spirit. The moot-point -far from being answered- is: to what extent the current education system reproduces the basis of the status quo or gives fresh opportunities to vulnerable students.

2.4 Private vs. Public Universities

Detractors and supporters of public education as well as public (state) university have not reached a consensus to date (Lieberman, 1989). The discrepancy has given by the intervention of the state to absorb the costs of public education or the benefits to promote the common wellness remains open in the academic circles (West, 1964; Levy, 1986). Over the recent decades, some voices alert that the withdrawal of government from direct founding is certainly based on ideological issues (Tilly, 1998). As John Holmwood (2011a) eloquently highlights, the accelerated expansion of the system evinced the rise of some pitfalls in modern education which include the students` concern about their future in the labour marketplace mainly marked with insecure and temporal jobs. To put the same in other terms, the commoditization of education becomes students (who are future researchers) into mere consumers. As the authors remark, the private sector has different levels of investment and funds in which case there is a type of strong hierarchy among universities. What is more salient, in the liberal market workers are paid according to their degrees and their formal education. In the US and part of Europe, blue-collar worker families have fewer possibilities to enter the higher education system than other groups. This, doubtless, creates a considerable asymmetry in the salary levels. Paradoxically, families are forced to incur long-lasting indebtedness to cover students` university costs. In consequence, universities act as real gatekeepers which legitimate the capitalist financial inter-class asymmetries. Notwithstanding, things come worse to worst, public funds invested by governments seem to fall short to reverse the inter-class inequalities. This suggests that more access to higher education does not equate to the system sustainability. Holmwood reviews the original function of public universities alerting that the promotion of some liberal market mechanisms has altered not only the performance but the goals of the public university. Designed chiefly to reduce the economic inequalities culturally enrooted in society, the public university was prone to include a diversity of social groups but gradually became in an ivory tower strictly reserved by an elite. Based on Kerr's term multiversity, which connotes the different functions of the university to enhance knowledge democratization, Holmwood holds that under some conditions this multiplicity of visions and voices invariably leads to social conflict (Holmwood, 2011b). Some studies based on mix-balanced argumentations stress that it is hard to enumerate the advantages or disadvantages of both public and private universities without considering the particular environment as well as the students´ satisfaction in both institutions. Even though public universities evince little engagement and autonomy regarding the main administration no less true is that private institutions' professional researchers claim to be subject to a climate of ongoing competition (under the logic of publishing or perish) and psychological distress. It is difficult to resist the impression that job satisfaction in campuses varies in university and social contexts regardless of its nature as public or private (Volkwein & Parmley, 2000; Tai & Chuang, 2014). As Foray & Lissoni (2010) adhere, the figure of university occupies a central place in the configuration of scientific research. With direct interaction with industry, scientists often devote considerable efforts to resolve the real problems of the private sector or the economic corporations they are ultimately recruited. The symbiosis between the private sector and university arises some much deeper issues. Farfetched as it may sound, there is a tension between the basic research activities –which are centred on the need to make the obtained outcome open- and the opportunities to commercialize academic inventions. Secondly and most importantly, private sectors offer solid intellectual rights which are fitted against the free accessibility and replication proper of science. In sum, the authors conclude that the introduction of alternative institutions to mediate between the private and public sectors is at least necessary. It is safe to say both (private and public) universities operate at different levels absorbing and solving those systemic pitfalls given by the democratization of higher education. Having said this, the globalisation of a knowledge-centred economy is shifting globally the nature and real function of universities. Disseminating the knowledge through the private industry, universities allow a rapid regeneration optimizing efficient driven economic plans (Hagen, 2002). Lastly, senior sociologist Michael Burawoy (2011) dissects the current state of crisis in applied research and the university. Per his standpoint, the golden age of university where professors lived in a type of ivory tower has withered away forever. Today scholars have no opportunity to engage with "a wider society" simply because they are subject to new (strict) regulations, rules and protocols resulting from the commoditization of higher education. As he notably writes: "*we face numerous pressures of instrumentalization, turning the university into a means for someone else´ end. These pressures come in two forms- commoditisation and regulation.* (Burawoy, 2011: 17).

But as clarified, it is not an overnight process; the capitalist nation-state is gradually withdrawing funds from higher education over decades leaving a gap that is fulfilled by private corporations. The old dilemma between public vs. Private education has short-fallen. Nowadays, the private sector not only monopolizes knowledge production but also set the agenda of higher education worldwide. The commodification model, which includes the invention of metrics and scholars or university rankings that instrumentalises all dimensions of higher education, should be coupled to a regulation model which activates a set of new rules and protocols oriented to protect the incursion of the private sector in universities. This model encompasses a set of governmental regulations and evaluations to validate Professors with a firm tenure certificate (track) at the university. In the next section, we shall discuss those issues in higher education triggered by the encroachment of the liberal market. Whilst the commoditisation puts the question on of knowledge production for protecting particular interests, regulation moulds the reason for knowledge production, which means what is an object of research and what is not.

2.5 Current Issues in Higher Education

The current problems in higher education are manifold, but very well they can be classified in two clear-cut trends. On one hand, the problems and limitations in teaching students in universities or institutes which mainly associate with gun violence and the urgency to keep universities safe, the decline of the critical thinking and academic performance in students, the abuse of alcohol and drugs, as well as the budget cuts which leads to further tuition fee and lower-wages at the staff without mentioning high dropout rates in pre-graduate students (only to name a few) (Burton & Bartlett, 2009; Parsons & Taylor, 2011). On another, serious limitations strictly applied to research that affect daily professional fieldworkers. To wit, the publish or perish culture which seems to be organised around two main index and editorial corporations Elsevier (Scopus) and ISI Thompson (Web of Science) place many scholars between the wall and the deep blue sea. In the US as well as part of Europe many Professors are pressed to publish in high-impact factor journals to keep their academic tenure (Noble, 1989). In other cases, academicians enter in a climate of extreme competition leaving the teaching to a secondary position (Bretag, 2012). In a seminal paper published in *the Journal of Management Inquiry,* De Rond and Miller (2005) question to what extent the existent pressures for publishing to keep the academic standards and metrics is a bane or a real boon. As they notably observe, the phenomenon seems to be feared by some whilst venerated by others; to say above all, in universities linked to business and management. Doubtless, this mandate has serious and direct effects on the profession whilst hurting the researcher's creativity. As they go on to say: "*publish or perish signifies the principle according to which a faculty member's tenure is principally a function of his or her success in publishing. It comprises a race against time that typically begins when the faculty member is hired and ends when the tenure decision.*

must be made" (de Rond & Miller, 2005: 322).

As the previous backdrop, de Rond & Miller argue convincingly that there is a clear risk in prioritising academic productivity instead of innovation and creativity. One of the main problems, per these specialists, lies in the fact the best researchers seduced to be paid for their publications opt to leave teaching to lower-ranked ones. The act of publishing ensures faculty members better positions, recognition and reputation bargaining power in a global ethos whereas teaching is reserved to be esteemed locally. This model not only is accelerated by the business school model but also with the arrival of the market in the higher education system. At a first hand, it behooves us to consider that the real problems in teaching (such as the lack of empathy between professors and students, as well as problems in professors and students´ motivations) should be traced back to the business school model (BSM). Secondly and most importantly, the decline in creativity widely supported by BSM leads to giving ad-hoc answers to the real problems society often faces. Ultimately, the lack of motivation impacts negatively not only in professors who are low paid but also in students showing an increase in the academic dropout rates. Besides, students who originally enrolled at the university maintain higher ideals regarding their future that do not necessarily meet with their presence at their labour market -once their degree is earned-. This point brings us back to the fact that students are daily subject to sentiments of great frustration and fear during their careers (Gulzar et al., 2012; Sarrico & Melo, 2012; Trammell & LaForge, 2017). In consonance with this, some authors call attention to making universities more entrepreneurial projects. As long as the process takes room, experts should evaluate those facilitators and barriers to bolster a fluid dialogue with the community (Kirby et al., 2011; Shane, 2004). For example, Yusof and Jain (2010) conduct an interesting review of the specialised literature finding three main categories in the process which comprises entrepreneurial university, academic entrepreneurship and university technology transfer. In 1980, the government passes the enactment of the Patent and Trademark Law Amendment Act (known as Bayh-Dole Act) which means the radical shift of all intellectual property policy to give incentive to researchers to make their investigation public. The goal of this act associates not only with the need to commercialise research but also suspend federal funding whilst putting it in the hands of the private market. This action prompted academic revolutions. The first revolution was given by the urgency to adopt applied research to resolve the glitches of the market. Secondly, a second revolution is aimed at integrating applied research with a clear enhancement of teaching and learning skills. A scan of the literature indicates that the term entrepreneurial university denotes those institutions which are critical to the local economic development. At the same time, academic entrepreneurship centres on creating new opportunities for businesses that help the university to engage with the commercialization of rights, research and technology transfer and other activities. In addition, the technology transfer from university to private sector activates a set of strategic opportunities to deal with the challenges posed on society (Wright et al., 2004; Morris et al., 2013). One of the global risks that placed higher education in jeopardy was the COVID19 pandemic originally reported in Wuhan China at the end of 2019.

2.6 The COVID-19 Pandemic and Higher Education

It is not surprising to acknowledge that nothing will be ever the same just after the COVID-19 pandemic. The SARSCOv2, well-known as COVID-19, was originally reported in Wuhan China but in weeks it spread to Europe and the US. To date, statistics report almost more than five million deaths whilst more than 400 million are infected. To stop the infection, the different governments disposed of desperate efforts as the closure of borders and airspaces, as well as the imposition of strictest lockdowns and social distancing. Schools, universities and any public buildings were automatically closed for months or years worldwide. Without any doubt, COVID-19 harmed the higher education system making more acute some imbalances and asymmetries. The pandemic affected the most vulnerable learners the hardest but, at the same time, digital technologies helped to potentiate online teaching in all levels of higher education (Burki, 2020; Rashid & Yadav, 2020). Crawford et al. (2020) share the outcomes of their investigation describing how COVID19 impacts different higher education systems across the world. Based on the examples and responses against the COVID19 widely documented in 20 countries, these researchers acknowledge that reactions of higher education providers have been diverse in the time and social context. Whereas some institutions showed no response to the social distancing in campuses others adapted faster to online teaching redrawing the curricula and syllabuses to offer a collective response. Those learners who had poor connection or accessibility to digital technologies experienced further dropout rates than other groups. In short, developing or underdeveloped nations with low accessibility to technology had fewer opportunities to frame higher education to the pandemic than the developed economies. Whatever the case may be, under some conditions these specialists conclude, digital technologies boost online learning mitigating the impacts of COVID-19 in higher education. As stated, COVID-19 has certainly changed universities and life on campuses in many ways. Bhagat and Kim (2020) accept not only COVID-19 challenge higher education globally as never before but also precipitated serious changes in the curricula. Per their stance, these transitional transformations towards online teaching are far from being useful for students. The fact was that 1.5 billion students were suddenly forced to abandon face-to-face learning to adopt remote learning. To some extent, experts misjudge the nature and the limitations of online learning. This simply happens because not all students are fit to access remote learning. At the same time, in online services, the students' academic performances seem to be in decline. As other studies have shown, Bhagat & Kim acknowledge that institutions dealing with disadvantaged or frustrated students have serious problems keeping engagement during the pandemic.

> *Another peculiar change stems from the concerns about the quality of online education. The higher education sector's movement toward remote learning has not raised questions about the value of online learning due to the limited or inadequate access to technology but it also raised a myriad of concerns about the quality of online teaching. A recent report indicates that only 15% of college students were enrolled in full-fledged online programs even before pandemic* (Bhagat & Kim, 2020: 367).

Another added problem associated with the financial deficits private and public institutions went through the pandemic as well as the rise of unemployment in the labour market. Thousands of families were systematically pressed to live in marginal conditions or impoverished because of the quarantine and the different imposed restrictive measures to global trade. Sub service sectors, tourism and hospitality industries were on the brink of collapse given the containing measures. Lastly, the literature suggests that students and professors experienced higher levels of anxiety and frustrations as long as the COVID-19 pandemic. Resiliency in higher education plays a leading role in adapting the educational infrastructure to the new emerging contexts which are mainly characterized by uncertainness and decentralised planning. Even if COVID-19 jolted the higher education system from any lingering complacency no less true is that digital technology mediates between students and professors and educational institutions in contexts of lockdowns and social distancing. These obtained outcomes coincide with other studies applied in different countries and social backgrounds. Experts valorise digital technology to cope with the challenges posed by the pandemic but they alert that technology does not suffice to reduce the dropout rates of students debarred from the accessibility to digital technology (Almazova et al., 2020; Marinoni et al., 2020; Agasisti & Soncin, 2021).

Last but not least, some worker unions in higher education overtly rejected the possibility to return to in-person classes once governments removed the restrictive measures. In many countries of Latin America, like Argentina where worker unions keep strong pressure to negotiate with the central administration, put a set of demands and conditions to adopt face-to-face tutorials during the pandemic which included the COVID-19 vaccination, the respect of health protocols as well as other restrictive measures at classrooms. As Daniela Perrotta puts it, Argentina is mainly marked by a much deeper tradition where public education takes the lead subordinating the private sector. State universities are self-regulated and governed institutions where there is a co-leadership among teachers, students, graduates and administrative staff (only to name a few). The curricula –far from being homogenized in the public university- are based on heterogeneous institutional negotiations and designs; what is more important, public universities centralise and monopolise (through public funds) a whole portion of scientific knowledge production in the country. Although universities served well in the fight against the pandemic engaging with scientific research as well as health authorities, the teaching activities in other disciplines and academic fields were seriously harmed. The answers of public universities to online teaching were certainly poor in comparison to other countries which had previous experiences in these types of pandemics. There is a clear inequality gap which exhibits previous conditions of vulnerability in lower classes that affected their engagement with the public university. Unlike other countries, distance education fails to modernize the existent structure; rather, it reproduces previous material asymmetries created by liberal capitalism (Perrotta, 2021). Though poorly documented, this moot point represents a commonplace of convergence for future debates in Latin America higher education. Hence the present book and PANCOE experiment situate as a successful instrument which fills the gap, innovated in modern pedagogy.

2.7 Conclusion

The present book chapter discussed in depth the current problems of higher education in the world. Although the main guidelines and debated and well-thought-out ideas should be implicated contextually since problems vary on the country and socio-economic background, no less true seems to be that higher education faces an unparalleled crisis since it was widely standardised and democratised just after the end of WWII. This crisis witnesses the lack of motivation in students and professors as well as the fund deficits in public universities. The 70s and 80s were mainly characterised by an economic stage of expansion which included other relegated groups and classes. Higher education, which was originally based on the need of protecting ruling class interests, was re-organised to encompass a broader spectre of new students. As a result of this, there was a clear tension between the public and private universities. The arrival of the liberal market not only accelerated the labour conditions but also changed the basic pillars of university. Education has been commoditized as a product to be sold to those classes which can pay for it. With the benefits of hindsight, researchers have been pushed to leave their classes to focus strictly on scientific research. The introduction of metrics and academic rankings closed the doors of a new design oriented to impose the mandate of publishing or perishing. As a result of this, professional research was globally valorised and well-paid leaving the act of teaching in a marginal (secondary) position. The crisis of higher education which comprises violence at universities, low-paid salaries, academic bureaucracy as well as the subordination of public institutions to the labour market, has been cemented by the rise of a new virus known as COVID-19. Having said this, COVID-19 not only changed dramatically the higher education system but also paved the way for new curricula designed based on online teaching. This poses the dilemmas of those actors who are relegated from the economic production system and dispossessed of the accessibility to digital technologies. The gap between developed and underdeveloped economies has been placed into the foreground because of COVID-19. In the next book chapter, we review the problems and challenges posed by over-tourism education in different parts of the world as well as its intersection to COVID-19.

References

Agasisti, T., & Soncin, M. (2021). Higher education in troubled times: On the impact of Covid-19 in Italy. *Studies in Higher Education, 46*(1), 86–95.

Almazova, N., Krylova, E., Rubtsova, A., & Odinokaya, M. (2020). Challenges and opportunities for Russian higher education amid COVID-19: Teachers' perspective. *Education Sciences, 10*(12), 368.

Benavot, A., Cha, Y. K., Kamens, D., Meyer, J. W., & Wong, S. Y. (1991). Knowledge for the masses: World models and national curricula, 1920-1986. *American Sociological Review, 56*, 85–100.

Bhagat, S., & Kim, D. J. (2020). Higher education amidst COVID-19: Challenges and silver lining. *Information Systems Management, 37*(4), 366–371.
Borum, R., Cornell, D. G., Modzeleski, W., & Jimerson, S. R. (2010). What can be done about school shootings? A review of the evidence. *Educational Researcher, 39*(1), 27–37.
Bretag, T. (2012). *Publish or perish: Ramifications for online academic publishing*. Emerald Group Publishing Limited.
Burawoy, M. (2011). Redefining the public university: Global and national contexts. In J. Holmwoord (Ed.), *A manifesto for the public university* (pp. 27–41). Bloomsbury.
Burki, T. K. (2020). COVID-19: Consequences for higher education. *The Lancet Oncology, 21*(6), 758.
Burton, D., & Bartlett, S. (2009). *Key issues for education researchers*. Sage.
Christopherson, S., Gertler, M., & Gray, M. (2014). Universities in crisis. *Cambridge Journal of Regions, Economy and Society, 7*(2), 209–215.
Crawford, J., Butler-Henderson, K., Rudolph, J., Malkawi, B., Glowatz, M., Burton, R., et al. (2020). COVID-19: 20 countries' higher education intra-period digital pedagogy responses. *Journal of Applied Learning & Teaching, 3*(1), 1–20.
de Ridder-Symoens, H., & Rüegg, W. (Eds.). (2003). *A history of the university in Europe: Volume 1, universities in the middle ages* (Vol. 1). Cambridge.
De Rond, M., & Miller, A. N. (2005). Publish or perish: Bane or boon of academic life? *Journal of Management Inquiry, 14*(4), 321–329.
Doti, J. L. (2004). Is higher education becoming a commodity? *Journal of Higher Education Policy and Management, 26*(3), 363–369.
Elias, N., & Dunning, E. (1986). *Quest for excitement. Sport and leisure in the civilizing process*. Basil Blackwell.
Foray, D., & Lissoni, F. (2010). University research and public–private interaction. *In Handbook of the Economics of Innovation, 1*(1), 275–314.
Gillespie, N. A., Walsh, M. H. W. A., Winefield, A. H., Dua, J., & Stough, C. (2001). Occupational stress in universities: Staff perceptions of the causes, consequences and moderators of stress. *Work & Stress, 15*(1), 53–72.
Gulzar, S., Yahya, F., Nauman, M., Mir, Z., & Mujahid, S. H. (2012). Frustration among university students in Pakistan. *International Research Journal of Social Sciences, 1*(4), 7–15.
Hadden, K., & London, B. (1997). *Educating girls in the third world: The demographic, basic needs, and economic benefits. Education in comparative perspective* (pp. 31–46). Brill.
Hagen, R. (2002). Globalization, university transformation and economic regeneration: A UK case study of public/private sector partnership. *International Journal of Public Sector Management, 15*(3), 204–218.
Holmwood, J. (2011a). Introduction. In J. Holmwoord (Ed.), *A manifesto for the public university* (pp. 01–11). Bloomsbury.
Holmwood, J. (2011b). The idea of a public university. In J. Holmwoord (Ed.), *A manifesto for the public university* (pp. 12–26). Bloomsbury.
K12 School Shooting Database. (2021). *Center for Homeland Defense and Security*. Available at https://www.chds.us/ssdb/data-map/
Kamens, D. H., Meyer, J. W., & Benavot, A. (1996). Worldwide patterns in academic secondary education curricula. *Comparative Education Review, 40*(2), 116–138.
Kirby, D. A., Guerrero, M., & Urbano, D. (2011). Making universities more entrepreneurial: Development of a model. *Canadian Journal of Administrative Sciences/Revue Canadienne des Sciences de l'Administration, 28*(3), 302–316.
Lawton, D., & Gordon, P. (2005). *A history of Western educational ideas*. Woburn Press.
Levy, D. C. (1986). 'Private' and 'public': Analysis amid ambiguity in higher education. In D. Levy (Ed.), *Private education* (pp. 1–15). Oxford University Press.
Lieberman, M. (1989). *Privatization and educational choice*. Education Policy Institute.
Mandler, P. (2020). *The crisis of the meritocracy: Britain's transition to mass education since the second world war*. Oxford University Press.

Marinoni, G., Van't Land, H., & Jensen, T. (2020). The impact of Covid-19 on higher education around the world. *IAU global survey report, 23.*

Mayhew, K. (2017). UK higher education and Brexit. *Oxford Review of Economic Policy, 33*(suppl_1), S155–S161.

McKay, S., & Rowlingson, K. (2011). The religion of inequality. In J. Holmwoord (Ed.), *A manifesto for the public university* (pp. 90–111). Bloomsbury.

Meyer, J. W., & Schofer, E. (2007). The university in Europe and the world: Twentieth century expansion. In G. Krucken, A. Kosmutzky, & M. Torka (Eds.), *Towards a multiversity* (pp. 45–62). Transcript Verlag.

Mickelson, R. A. (2003). When are racial disparities in education the result of racial discrimination? A social science perspective. *Teachers College Record, 105*(6), 1052–1086.

Miller, N., & Sabapathy, J. (2011). Open universities: A vision for the public university in the twenty-first century. In J. Holmwood (Ed.), *A manifesto for the public university* (pp. 42–55). Bloomsbury.

Morris, M. H., Kuratko, D. F., & Cornwall, J. R. (2013). *Entrepreneurship programs and the modern university.* Edward Elgar Publishing.

Müller, S., & Schneider, T. (2013). Educational pathways and dropout from higher education in Germany. *Longitudinal and Life Course Studies, 4*(3), 218–241.

Noble, K. A. (1989). Publish or perish: What 23 journal editors have to say. *Studies in Higher Education, 14*(1), 97–102.

Parsons, J., & Taylor, L. (2011). Improving student engagement. *Current Issues in Education, 14*(1), 1–19.

Perrotta, D. (2021). Universities and Covid-19 in Argentina: From community engagement to regulation. *Studies in Higher Education, 46*(1), 30–43.

Piché, P. G. (2015). Institutional diversity and funding universities in Ontario: Is there a link? *Journal of Higher Education Policy and Management, 37*(1), 52–68.

Rashid, S., & Yadav, S. S. (2020). Impact of Covid-19 pandemic on higher education and research. *Indian Journal of Human Development, 14*(2), 340–343.

Reay, D. (2012). Universities and the reproduction of inequality. In J. Holmwoord (Ed.), *A manifesto for the public university* (pp. 112–126). Bloomsbury.

Rojek, C. (1993). *Ways of escape: Modern transformations in leisure and travel.* Springer.

Rojek, C. (2013). *Capitalism and leisure theory (Routledge revivals).* Routledge.

Sadlek, G. M. (2014). Otium, Negotium, and the fear of acedia in the writings of England's late medieval Ricardian poets. In M. Fludernik & M. Nandi (Eds.), *Idleness, indolence and leisure in English literature* (pp. 17–39). Palgrave Macmillan.

Sarrico, C. S., & Melo, A. I. (2012). Let the devil choose: Frustration or anxiety in the wake of performance measurement in universities. In B. Stensaker, J. Valimaa, & C. Sarrico (Eds.), *Managing reform in universities* (pp. 81–97). Palgrave Macmillan.

Scott, J. C. (2006). The mission of the university: Medieval to postmodern transformations. *The Journal of Higher Education, 77*(1), 1–39.

Shane, S. A. (2004). *Academic entrepreneurship: University spinoffs and wealth creation.* Edward Elgar Publishing.

Sierpinska, A., Bobos, G., & Knipping, C. (2008). Sources of students' frustration in pre-university level, prerequisite mathematics courses. *Instructional Science, 36*(4), 289–320.

Stebbins, R. A. (2016). Dumazedier, the serious leisure perspective, and leisure in Brazil. *World Leisure Journal, 58*(3), 151–162.

Tai, F. M., & Chuang, P. Y. (2014). Job satisfaction of university staff. *The Journal of Human Resource and Adult Learning, 10*(1), 51–63.

Tilak, J. B. (2008). Higher education: A public good or a commodity for trade? *Prospects, 38*(4), 449–466.

Tilly, C. (1998). *Durable inequality.* University of California Press.

Tinto, V. (1975). Dropout from higher education: A theoretical synthesis of recent research. *Review of Educational Research, 45*(1), 89–125.

Tomlinson, S. (2019). *Education and race from empire to Brexit*. Policy Press.
Trammell, B. A., & LaForge, C. (2017). Common challenges for instructors in large online courses: Strategies to mitigate student and instructor frustration. *Journal of Educators Online, 14*(1), 1–15.
Veal, A. J. (2018). *Whatever happened to the leisure society?* Routledge.
Veblen, T. (2017). *The theory of the leisure class*. Routledge.
Volkwein, J. F., & Parmley, K. (2000). Comparing administrative satisfaction in public and private universities. *Research in Higher Education, 41*(1), 95–116.
West, E. G. (1964). Private versus public education: A classical economic dispute. *Journal of Political Economy, 72*(5), 465–475.
Wike, T. L., & Fraser, M. W. (2009). School shootings: Making sense of the senseless. *Aggression and Violent Behavior, 14*(3), 162–169.
Wilbrink, B. (1997). Assessment in historical perspective. *Studies in Educational Evaluation, 23*(1), 31–48.
Witmer, D. R. (1970). Economic benefits of college education. *Review of Educational Research, 40*(4), 511–523.
Wright, M., Birley, S., & Mosey, S. (2004). Entrepreneurship and university technology transfer. *The Journal of Technology Transfer, 29*(3), 235–246.
Yusof, M., & Jain, K. K. (2010). Categories of university-level entrepreneurship: A literature survey. *International Entrepreneurship and Management Journal, 6*(1), 81–96.
Zuzanek, J. (2020). *Time, leisure and well-being*. Routledge.

Chapter 3
Tourism Education in Perspective

Abstract To some extent, tourism education is facing an unparalleled crisis, probably a terminal crisis. Metrics applied in research and publications are wreaking havoc in the discipline as never before. Many professional researchers are pressed to publish leaving their classrooms to amateur professors. This chapter insights on the genesis and evolution of higher tourism education as well as the curricula formation. At a first glimpse, the indiscipline of tourism research -originally denounced by Tribe- is accompanied by other problems which include high dropout rates, a dissociation between professional research and learning without mentioning low salaries and bad working conditions once the Ba is earned. The present chapter interrogates the present problems of higher tourism education and the future of the discipline in an ever-changing and global world.

Keywords Education · Emancipation · Dropout rates · Motivation · Higher tourism education

3.1 Introduction

Tourism research has its roots in the well-functioning of the industry as well as in tourism management. The specialised literature punctuates that some global dangers have placed the industry in jeopardy. To wit, policymakers have emphasised the importance -if not urgency- to update the tourism curricula to the real problems, risks and changes tourism is facing today (Cunliffe, 2002). These risks encompass terrorism, climate change, and resiliency -only to name a few (Du, 2003; Jafari & Ritchie, 1981; Sheldon et al., 2011). In a seminal paperwork, Cooper and Shepherd (1997) argue convincingly that the intersection of tourism education and the industry is very complex –if not mainly marked by a set of barriers and limitations. Both scholarship and the industry can share mutual benefits whilst working together-. Tourism education centres on the delivery of stakeholders´ expectancies –unlike other disciplines such as sociology or geography-. In this vein, Amoah & Baum (1997) lament on the fact not always universities or curricula are aligned with the

A. Zuccoli, M. E. Korstanje, *The Role of Pleasure to Improve Tourism Education*, https://doi.org/10.1007/978-3-031-21580-3_3

national policies in the fields of tourism. Airey and Tribe (2006) acknowledge that the rapid growth of studies and publications does not coincide with the maturation of the discipline. There is an underlying tension between applied research and the managerial perspective which over-valorises profitability and management as valuable instruments towards development. Knowledge production is mainly determined by a type of indiscipline where the methodological (if not epistemological) basis of tourism is far from being clear. This invariably impacts the curricula as well as the training of next workforce. Authors call attention to the importance to adopt new methodologies to enhance the curricula in tourism education. In the mid of this discussion, some pungent questions simply surface: what are the main goals and guidelines of tourism curricula? should be tourism education focused to resolve the problems of the industry? what are the main pitfalls is facing tourism education today? Is tourism education in crisis? if so, why?

All these questions do not have a unique and simple answer, albeit problems in tourism education seem not to be pretty different than problems in higher education –a point well-debated in earlier sections. As the first entry in this discussion, we might speculate that Cartesian dualism gravitates seriously in the formation of educators and curricula in the constellations of tourism. Cartesian dualism has monopolised the curricula content whereas imposing quantitative-led methods and positive reasoning. The outset of tourism education was interlinked to the urgency to find those threats that affect the integrity of tourist system. In this token, educators emphasised on metrics and mathematical algorithms which assisted them to validate or refute their original hypotheses. The consolidation of tourism -as a maturated discipline- historically depended upon its capacity to measure (not describe) global trends. As we shall discuss in the present chapter, this academic position led invariably to the hegemony of what scholars dubbed as *economic-centred paradigm.* This doctrine gave priority to profitability and management over other social issues. To put the same in brutally, the mandate of the industry first, the scientific research later! comes to stay in tourism education. The first section of the present book chapter interrogates on the underlying factors that facilitated the genesis of tourism education. From its inception, the pioneers of the discipline have been mainly concerned by the future of tourism research as well as the gap between theory and practice. One of the main questions that guided their work has been associated to the education of future workforce in an ever-changing world. Complementarily, the second section signals to the rise of three clear academic traditions. At the same time, we put each one on the critical lens of scrutiny. The epistemological tradition struggled to find an all-encompassing definition of tourism. Rather, the economic-centred tradition focused on the management and profits of international destinations. Per this academic tradition, the tourist is the main epistemological source of knowledge. Later the critical tradition or critical turn devoted considerable time in studying the negative effects of the industry in society. The third section introduces a hot debate revolving around the dissociation of teaching and researching. The culture of publishing or perish is wreaking havoc in the higher education system. Scholars are often pressed to publish their advances leaving the classrooms to less experimented colleagues. In the fourth section we examine the future of gamification and the concept of the flow, which is strictly inscribed into the state of well-being.

3.2 The Genesis of Tourism Education

One of the pioneering voices in tourism education Jafari and Ritchie (1981) alert that there is a weak dialogue between the industry and the resulted curricula development. Per their viewpoint, policymakers should identify the dynamic and interaction among the different stakeholders such as visitors, goods, and infrastructure as well as host-guest relationships. For instance, Jafari, in his prolific career, has advocated for tourism education development as well as the formation of students´ curricula. Preliminarily, he discerns four clear cut platforms: advocacy, cautionary, adapting and knowledge-based platforms. Each platform indicates not only the direction of research but also the evolution of tourism in the threshold of time. At a first glimpse, *the advocacy platform* is certainly formed by a set of studies and works praising the economic benefits of the tourism industry and its positive impact on society. At the time, other critical works focused on the social maladies accelerated by the adoption of mass tourism. These studies comprise what Jafari called "the cautionary platform". Jafari and his colleagues strongly believed that the evolution of tourism research (education) is given by the number of published texts, books and papers located in leading journals. The adaptancy platform exhibits a mix-balanced diagnosis where specialists report the advantages and disadvantages of the industry. The knowledge-based platform marks a new stage of maturation where research centres strictly on observable (validated and tested) facts (Jafari, 2001, 2002). Over years, Jafari was an authoritative voice who indisputably shed light on the constellations of tourism research and education (Hall et al., 2016; Xiao, 2013). Given the problem in these terms, Cody Paris (2011) observes that one of Jafari's contributions illuminated on the dualism of the tourism industry. This dualism rests on the fact that tourism is approached as a social phenomenon (widely studied by social sciences) and a business-related activity (addressed by managerial disciplines). Doubtless, tourism education not only follows this dualistic nature –as Tribe (2008) brilliantly discusses- but also educates the next generation to deal with and resolve the global challenges that place the industry in jeopardy (Paris, 2011). Coming back to this point, as John Tribe (2008) recognizes, the dualism in tourism higher education seems to be an inexpugnable gap that cannot be currently filled by experts. Critical studies alert on the risks of adopting economic-based paradigms in applied research whereas policymakers blame social scientists to work with hypotheses that fall short to fix the real problems of the industry. Going in a balanced direction, Tribe postulates the thesis that critical studies help in creating more ethical management without neglecting employability skills. As he writes, *"critical theory can subject conventional management plans to ideology critique and can offer a way of generating a more meaningful stakeholder analysis and one where other the aspirations of "other" stakeholders are not subjected to an ideological framing and interpretation by the most powerful"* (Tribe, 2008: 253).

As the previous argument is given, Tribe toys with the belief that tourism education offers a fertile ground to cultivate critical thinking enhancing not only the curricula but also incorporating critical studies towards a more sustainable form of commercial exploitation (Airey et al., 2015; Korstanje, 2022; Tribe, 2008). At its beginning, tourism education -probably moving as a sub-discipline of tourism research- focused on the question of sustainability and governance in the curricula formation. It is important to note that Knowledge production, which is re-channelled through education, occupies a central position in improving local governance and political stability (Coles, 2009; Dredge, 2015; Moscardo, 2011). This begs some vexed points: does tourism education suffice to make more resilient destinations in a world fraught with risks and uncertain conditions? What are the limitations of the above-noted dualism to deal with these new conditions just after the turn of the century?

In this respect, Sheldon et al. (2011) use the term TEFI (Tourism Education Futures Initiatives) to denote the provision of a long-term vision and theoretical framework to be incorporated in tourism educational programs. This model lays fingers not only on educating citizens for a better (global world) but also the need to educate the workforce to manage the problems of the industry in the complex years to come. TEFI includes foundational values in tourism education mainly characterised by ethics, stewardships, knowledge, mutuality and professionalism. Whereas ethics is esteemed as the touchstone for a good action, or common good, stewardship corresponds with the idea of responsibility and reciprocity. At the same time, knowledge is mainly based on the necessary skills and expertise the person acquires through the education process. Mutuality and professionalism speak to us of those skills and competencies given for staff to be entwined to others like the guest or the customers who need our assistance. Problems of tourism education aggravates in view of the expansion of globalisation and the interests of Asian students for visiting Europe and the US. As never before education should homogenise global (overseas) students all them with different cultural background, traditions, hopes and fears. Federal and local governments often confront to monopolize the resources derived from higher education in tourism and hospitality. In consonance with this, Don MacLaurin (2006) calls the attention on what specialists dubbed as "Ontario experience". In response on the urgency of recent professionalization of the tourism industry, in 1967 Provincial Government of Ontario made the decision to create 22 new community colleges. Central administration struggled to pass some laws to homogenise tourism higher education but without any result. This happened simply because Federal government in Canada does not set university tuition rates, but rather they are subject to local laws. At the same time, the private sector increased financial support for tourism and hospitality education. This was not the same example of the US widely described by Tom Baum (2011) who remarks on in a preliminary stage, the UK institutions were reluctant to accept substantial changes in the curricula. However, the expansion of the tourism industry as well as the internationalization of tourism demand introduced radical shifts in the UK universities which were adapted to over-seas students. As Baum observes, tourism schools internationalised their curricula so that the programs reflect the needs of graduates to move inside the ebbs and flows of an international tourism market not only at home but also abroad. This internationalization process includes the adoption of foreign languages, the elaboration of a new curriculum in collaboration of new arrangements with international schools in Europe and Australia (Baum, 2011). By this side, Arie Reichel (2011) brings an interesting reflection on the case of Israel. For 90s decade, universities in the country have no intention to encourage tourism careers internally. Managers and employers holding university academic degrees were scarce. The opposite was equally true, the tourism industry was not an attractive for students simply because it not provided profits and well-paid jobs for educated people. The internationalization of tourism happened in the 2000s decade opened the doors to more decentralised forms and less state-controlled programs. Without any doubt, as Reichel puts it, tourism interrogated furtherly the classic higher education system in Israel.

As the previous argument is given, Paul Fidgeon (2010) asks furtherly on the importance of the curricula for higher education in tourism. It is important not to lose sight of the fact that the growth of the tourism industry overlooked the necessity to review the curriculum at most important universities. A scan of the literature says that knowledge skills development, which acts as a catalyst that prepares students for the labour market, plays a vital role in higher tourism education. Paradoxically, this position silenced other voices which were pushed away to a marginal position. As Fidgeon eloquently notes, tourism education valorises profitability over employability misjudging that academic vocational and transferable skills whilst adding further value to the industry. Evoking the Australian case, Phillip Pearce (2011) recognizes that in the early days, where the industry was in its infancy, there was a desperate effort to recruit under-skilled staff some of them even with marginal knowledge on the activity. Tourism staff gradually experienced a substantial modification according to the creation of new international demands. The classic bachelors demanded to enter in the industry sets the pace to Ph doctorates. The new workforce was dotted with emerging disciplines such as psychology, geography, anthropology as well as the managerial perspective including management, planning and marketing. Unlike other countries (like the US and the UK), in Australia the industry offers well-paid jobs. The higher education in tourism still continues flourishing as a fertile ground for foreign students in basis of the selective competence in the labour marketplace. In fact, students often move in a climate of full employment and freedom once graduated giving the discipline a source of status and international recognition. However, in the recent years, government has some problems to assess internally the success of Australian programs as well as the efficacy of tourism research in view of the lack of homogeneity in educational institutions. The opportunism to get good jobs in the labour market varies on culture and country, as Pearce ultimately alerts. For example, Carlos Costa et al. (2016) highlight that beyond the panacea of tourism education lie serious material asymmetries and gender issues which need to be corrected. Although women represent the majority of workforce -in Portugal-, they are lower-paid in comparison to men. Such a situation raises not only ethical problems but also gender differences. However, per their outcome, there are more female and male occupying leading position in the tourism and hospitality industries. Authors go on t write: *"it was concluded that nowadays, women prevail in higher education and, more specifically, in tourism higher education. Despite that, several inequalities between men and women persist in the labor market (e.g. gender pay gap, lack of representation of women in upper-management positions, etc). Even though higher education has the potential to raise the expectation of women and lead them to search for jobs that suit their qualifications, several studies demonstrate that having higher levels of education does not lead to equality between male and female salaries"* (Costa et al., 2016: 208). What is equally alarming, a whole portion of women who duplicates the students' rates never are recruited in the labour market or simply fail to earn their degree. Under some conditions which is very hard to precise here, women occupy low-level positions (King et al., 2003). Gretzel and Bowser (2016) remind that having women in upper-management engaged with higher levels of organization power reduces gender segregation. This happens because female leaders often bring some substantial changes and benefits to organizational culture. Women in companies or organizations with few female leaderships have little possibilities to get support from their colleagues. As a social function mainly marked by solidarity, role models take the lead to insert women in equal conditions comparing to men. To cut the long story short, tourism education has advanced a lot in recent decades but chaotically and in a disordered way. Notwithstanding this fact, this growth was chaotic and disconnected. As John Tribe puts it, the indiscipline of tourism rests not only in the indifference of the Academia for regulating strictly the knowledge production but also by the rise of different tribes and islands which move freely without any type of interaction with other academic circles. As a result of this, the notable growth of tourism literature has invariably led to a state of disorder and epistemological fragmentation (Tribe, 1997, 2000, 2004, 2010). In the next section, we review in-depth some of the most important traditions and theories that gravitated in tourism education worldwide.

3.3 Theories and Traditions in Tourism Education

Before starting, it is important to mention that any classification or selection on tourism education theories is a task almost impossible to perform, if not so to speak arbitrary. This suggests that our selection may be partially subject to some critical approaches. Anyway, we have made a selection of the most cited and discussed texts which moulded the discipline from its inception. A historical sweep of tourism education signals an uncanny dualism between the learning process and what should be taught to cope with the challenges posed on the industry (Cooper & Shepherd, 1997; Inui et al., 2006). Of course, as Tribe probed, the indiscipline of tourism epistemology is based on a lack of lighthouse to follow at the time of reconstructing a map on tourism education theory. Again, like our selection, the criterion used to review the material is always personal and discretional. Bob McKercher (2002) warns on the snail's pace of tourism education to be situated as a consolidated discipline. Professor David Airey explains that tourism in the higher education system is far from being consolidated since it has almost no more than 40 years old; from the outset, the discipline has engulfed to be taught in different western establishments and colleges but borrowing contents and academic programs from diverse other disciplines as geography, psychology and economy. Here, we can distinguish three main traditions. It is difficult to resist the impression that the epistemological tradition seeks to find valid answers to the problem of tourism –not as an industry- but as an object of study. For these scholars, there are many tensions between different theories orchestrated to define the nature of tourism. The fast growth of the tourism industry, as well as mass education, prompted governments to create new tourism-based careers at western universities. The training of the next workforce to re-organize the industry just after the 60s decade paved the way for the rise of multiple courses, careers, and bachelors in the US and Europe. A new emerging tradition chiefly oriented to set the students´ curriculum coincided with the explosion of tourism research which reflected with the number of academic journals (almost 150 to date) created to deal with the contradictions and real problems of the industry. A third tradition known as the cultural turn interrogated directly the main asymmetries and intricacies generated by the industry in a hyper-globalised world. Works framing this academic family focus on how the introduction of *free trade* erodes not only the basis of the public university but also commoditises tourism education globally. These studies emphasise the need of revising the existent academic metrics and the doctrine of publishing for conserving tenure. To cut the long story short, scholars, who take part in this academic circle, seem to be certainly worried about the future of tourism education in the neoliberal market; above all the dissociation between teaching –an activity reserved for low-cited scholars- and researching –professionally conducted by high-cited scholars- (Airey, 2016). This aspect will be addressed in the last section of the current book chapter. Echoing Airey's contributions, we differentiate three academic traditions in tourism education, which are discussed in the next sections, the epistemological tradition, the economic-centred tradition and the critical tradition.

3.3.1 The Epistemological Tradition

Although the epistemological tradition has a long trajectory in tourism fields, it may be very well traced back to the birth of St. Gallen School. Economists like Walter Hunziker or psychologists like Jost Krippendorf devoted considerable resources in defining the nature of tourism as well as calibrating the epistemological borders of the discipline (Akoglu, 2015; Korstanje, 2015). In a trailblazing book entitles *Holiday Makers,* he argues convincingly that the human mind is based on contradictory necessities such as working, relaxing or duties whilst escaping. The need for evasion keeps the self-working integrated with other inner drives. Most certainly influenced by Freud's conception of homeostasis any need should be satisfied so that the self can restore equilibrium. In the same way, tourism –far from being a mere industry- catalyses to alleviate all daily frustrations for lay citizens. Tourism keeps society united. Having said this, it is safe to say that holidays –as a temporal break- are widely enthusiastically valorised in almost all cultures in the world. As Krippendorf notes, holidays should be considered as mere social construes designed not only to keep the society functioning but also to frame the citizen's consciousness. Travellers go out of the home to start a liminoid tri partite fragmentation: living-here, working-there, and relaxing-anywhere. In Krippendorf's argumentation, there is a natural tourist drive that leads us to abandon temporarily our homes in quest of something new. This so-called tourist drive emulates life without burdens, regulations or procedures. Put simply, travelling is esteemed as a sign of status. Those persons legally encouraged to travel abroad are situated on the top of the societal hierarchy. Whatever the case may be, as he laments, tourism is involved in some economic forces that alienate the modern man who is constrained by countless advertising campaigns and mass media. This occurs because modern society is obsessed to reach happiness leading citizens to a fabricated world characterised by depersonalization and hedonist consumption. In this stage, citizens become consumers, as Krippendorf concludes (Krippendorf, 2010). Without any doubt, the figure of escapement is of paramount importance to understanding Krippendorf's legacy. From his standpoint, tourism is neither good nor bad; it is intertwined with the mainstream cultural values of society. As a pioneering scholar worried by ecology, Krippendorf opens a conceptual bridge between the sphere of labour and leisure laying the foundations towards a new understanding of tourism and its connection with modern society (Korstanje, 2015). His works inspired a young brilliant mind who has illuminated the first steps of tourism epistemology, Neil Leiper. Unlike Krippendorf who delves into the constellations of tourist behaviour, Leiper is moved to understand the semantic origin of the term, tourism. He strongly believes that etymology is a useful instrument to reveal the roots and evolution of tourism. Leiper traces back the origin of tourism to the first Tour organised by a French family, De La Tour to bring British traders to France in the Eighteenth Century. This alliance is considered as a foundational event for the rise of shopping travellers, and of course for the institution of Grand Tour widely practised by the European aristocracy (Leiper, 1979, 1983). Although Leiper's efforts to unveil the nature of tourism, no less true is that its etymological origin is a matter of controversy. For example, the Anglo-Saxon School holds the thesis that the term tourism stems from the old Anglo-Saxon word *Torn* which means surrounding or going around. From *Torn,* we have two derived terms: *torn-are (the act of going around), and torn-us (those who are going around).* The word was often used by medieval farmers to connote expeditions or travels mainly marked by the need of returning (Fernandez Fuster, 1967; Korstanje & George, 2022). Additionally, Artur Houlot (1961) finds some correlation between tourism and the Old Abrahaamic term *Tur* employed for Moses' explorations to Canaan described in the Old Testament. The conceptual discrepancy revolving around the tourism etymology far from being solved is today aligned to a clear fragmentation about its definition and nature (Korstanje, 2007). It is noteworthy that the founding parents of tourism epistemology have certainly left a rich legacy and conceptual framework to understand tourism that unfortunately has not been continued by scholars (Korstanje & George, 2022). Instead, the epistemological tradition turned its attention to the opportunities of tourism to become a maturated discipline. This behooves us to consider the following controversial aspects of tourism research: what are the conditions for tourism to situate as a maturated and well-recognised discipline? Here the discussion goes through two opposite pathways. On one hand, scholars of the calibre of Michael Hall, Tim Coles and David Timothy Duval, toy with the belief that the consolidation of tourism research is given by the possibilities to adopt multidisciplinary (or citing them post-disciplinary) methodologies (Coles et al., 2006; Coles et al., 2009). On another, voices like John Tribe who writes about the "fragmentation of tourism research", a dispersion of the created knowledge where academicians certainly move in isles. Per Tribe, the lack of interest in the Academia to set a clear epistemological agenda, conjoined to the dispersion of knowledge production, resulted in clear indiscipline ushered researchers in a state of solipsism and confusion. The adoption of techniques and methods coming from other disciplines aggravated ultimately the situation, as Tribe laments (Airey & Tribe, 2006). Furthermore, tourism research has been influenced by the economic-centred paradigm with strict priority in profitability and business in place of critical approaches. As Tribe comments, the managerial gaze not only is monopolizing the publications in the leading journals but also has changed the curricula in higher education. The economic-centred paradigm –which is based on management and marketing-, is not interested in describing tourism as a scientific object of study but only as a product, or an industry. Studies taking part of this paradigm emphasise those policies and courses of action oriented to protect the tourist destination (Tribe, 1997, 2000, 2010). For Tribe, this position invariably impedes and delays discipline maturation (Korstanje & George, 2022). In a nutshell, a new line of tourism epistemologists has called attention to the urgency to re-build the basis of tourism epistemology incorporating the conceptual background given by the founding parents. In fact, they caution on the risks to throw the baby out of the bathwater (Belhassen & Caton, 2009; Botterill, 2001; Rakić & Chambers, 2012).

3.3.2 The Economic-Centred Tradition

Most certainly, the economic-based tradition has gravitated profoundly in the formation of higher education tourism curricula. Historically, the economic-centred tradition has emphasised the need of protecting the organic image of the destination as well as stimulate competition. In the process, education serves as an efficient instrument to train the next workforce. One of the main aspects that worry educators seems to be linked to the lack of updating in higher education curricula (Chon et al., 2020; Cooper, 2002). In perspective, the rise of global risks which include terrorism, natural disasters –if not the recently COVID-19 pandemic- place the tourism industry in jeopardy. Policymakers –and educators- should work hard to update the curricula according to the new times (Prebežac et al., 2016). As Larry Dwyer (2012) asserts, a set of different stakeholders and actors have boosted the economy of tourism globally. The complexity of the tourist system is certainly given not only by the diversity of actors but also external factors (dangers) which impact negatively the sector. Since no single driver will dominate the global scenario in the decades to come, no less true is that the economic growth of tourism is subject to moments of instability and uncertainness. To the environmental problems, one must add the increase of competition among destinations because of trade liberalisation. The success of the decision-making process depends directly upon the levels of integration among actors and stakeholders. Following Dwyer, the main challenge for the industry associates with the ability to transform the curricula whilst adopting better knowledge processing practices. At a closer look, educators should contemplate different sub-disciplines such as marketing, politics, management, safety and security issues, accounting-finances as well as market-related drives operating in the system. In this vein, Wang et al. (2010) bring some critical reflection on the importance of students´ curriculum to be calibrated to the necessities of policymakers and the tourism industry. Based on Australia as the main study case, these researchers conduct a survey on managers and front desk staff members to understand their expectations and perception about the requirements of the industry as the answers the higher education often provides with. As outstanding as it may sound, the lack of updating leads some managers to be reluctant to accept the value of university graduates. Authors suggest bolstering a frank dialogue between higher education and the referents of the industry. Because there are low levels of educational specificity, managers are less enthusiastic to welcome the significance of a tourism degree. What seems to be more important, there is a gap between theory and practice which merits to be debated. Whilst tourism practitioners over-valorise operational skills to make the job successfully, academics emphasise strategic kills to offer an all-encompassing diagnosis of the problems the industry goes through. Similar conclusions have been certainly reached in other studies such as Weber and Ladkin (2008); Ladkin (2014) & Hsu (2018). Not only the curricula but also pedagogy as well as program content and offering should be drastically changed to overcome the state of crises the industry is facing (Boley, 2011; Lam & Xiao, 2000). Despite the development of studies and applied research in the fields of tourism education, some critical voices have lamented the discipline remains bogged down of the conceptual premises of the economic-based paradigm. The fact lies that a whole portion of what is produced and published in the constellations of education are under the auspices of the economic-based theory. In this vein, educators are conditioned to the fluctuations and needs of the tourism sub-service sector. This is a point of departure widely discussed by the Critical Tradition.

3.3.3 The Critical Tradition

The critical tradition (turn) has placed the current definitions revolving around the nature of tourism into the critical lens of scrutiny (Ateljevic et al., 2005; Ateljevic et al., 2013). In consonance with this, John Tribe (2001) distinguishes three conceptual paradigms in tourism education; each one conserving differences and issues in the curriculum design. Tribe finds that the scientific-positive paradigm alludes to the concentration of information that can be empirically validated through research. Researchers who embrace this paradigm are well-versed in quantitative-led methodologies. Per this doctrine, tourism should be studied contemplating the principle of value neutrality –as a thing-. Over the recent years, some scholars have criticised the principle of value neutrality simply because it is at least incomplete. The ideological position of field workers, as well as some environmental factors, exerts considerable pressure and influence in the professional research. This opens the doors to the emergence of a new paradigm he dubbed as "the interpretative paradigm". For this tradition, tourism education is commoditised according to the positive imperatives. Students should engage with the social world and human interaction to understand the role played by the tourism industry in society. It is tempting to say that the curriculum should be designed, contested and negotiated taking an active part of all involved actors. As discussed, the critical paradigm denotes special interest for tourism world-making, whilst denouncing the inter-class asymmetries replicating daily through tourism curriculum. Resulting from the positive methodology, the tourism curriculum is subordinated to the economic-centred paradigm. As Tribe adamantly debates, this dissociation between the industry as well as the social factor gravitates directly in the dilemma of teaching vs. Researching. Neither professors nor postgraduate students have easy access to the circle of professional researchers. Students often come across several barriers and problems at the time to publish their doctorate advances. In view of this, Kathleen Caton –in different approaches- calls attention on the importance to imagine tourism as something else than a mere economic activity, but as a space of multicultural relations and interaction (Korstanje & George, 2022). The managerialistic model, as she overtly claims, has created a unilateral definition of tourism monopolizing the content of curricula in English speaking countries. Therefore, harsh international criticism exerted on positivism and the current curricula in higher education have been brought onto the foreground. The vocational insight has led educators to consider tourism as a business-related activity alone whilst excluding other voices and methodologies into the periphery. In consonance with Tribe, the critical turn holds that tourism should be defined as a world-making force that exhibits the counter of different actors, who are culturally enrooted in different biographies. For that, notions such as skills, or competencies should set the pace to new terms like wisdom, understanding, and of course critique (Belhassen & Caton, 2009; Caton, 2014; Stevens et al., 2019). Other studies coming from anthropology and ethnology have focused on the power of local cultures to set the agenda of educators in the next decades (Young & Maguire, 2017). It is not simplistic to say that the act of teaching and learning appears to be going beyond the paradigms of vocationalist programs. They are subject to complex power relations which need to be carefully deciphered. Students´ capabilities and skills should be centred on more integrated and sustainable forms of learning (Beckendorff & Zehrer, 2017). Farfetched as it sounds, some students are baffled by the pressures and constraints academicians have in their professional careers. More interested in publishing than teaching, professional researchers prioritise academic work over teaching hours. The liberal market has introduced some radical transformations that are changing the student-professor relationships worldwide. One might speculate that teaching has been marginally reserved by low-cited or low-ranked scholars whilst the most consolidated academicians lay their fingers on leading –top-ranked-journals. This moot point will be addressed in the next section.

3.4 Teaching vs. Researching

It is safe to admit that a multiplication of journals and guidelines to publish has globally mushroomed. Scholars envisaged that the discipline maturation was certainly given not only by the number of tourism-led journals but also by the citation impact factors (Korstanje, 2022). Having said this, several studies published the ranking of more prolific and cited scholars worldwide. The prestige and status of tourism research were mainly centred on the impact factor and journal metrical values (Benckendorff, 2009; Gursoy & Sandstrom, 2016; Law, 2017; Pechlaner et al., 2004; Ryan, 2005; Zhao & Ritchie, 2007). However, little attention was paid to the effects of metrics in the higher education system. The dichotomy between teaching and researching is far from being new. The issue was hotly debated in earlier chapters. Of course, tourism higher education is not an exception (Ertaş & Kozak, 2020; Yankholmes, 2014). Professional researchers are pressed to publish their advances and works to keep their tenure. As Michael Hall (2011) puts it, the problem and the obsession for publications aggravate with the adoption of metrics and bibliometric analysis to rank scholars. At a first glimpse, bibliometric analysis is significant to measure the citation impacts of tourism scholars in the fields as well as in other disciplines, but no less true is that the involved metrics are unilaterally set by governments and private organizations. As a result of this, knowledge production is externally drawn and standardised beyond the control of professional researchers. Similar constraints or pressures face Ph. D students to sustain their dissertation. Some instructions request students to publish some of their Ph Dissertation thesis in top-ranked journals so that their universities may be better positioned in the international indexed rankings. In this context, Neil Carr & Stu Hayes (2017) explain that PhD graduates should not have any pressure to publish –at least as they apply for a formal academic position. But frequently they are encouraged to present to publish as lead authors in top-tiered journals. Bob McKercher (2008) advices on the importance to contemplate other forms of measuring academic scholars than citation impact factors or metrics. Per his viewpoint, there is a great concentration of the academic elite about the published content. From 58 tourism journals registered, McKercher adds, 500 citations were reported since 1970. What is more important, from this cypher almost 250 citations are located in 48 authors only. The publish-or-perish paradigm has recently been criticised by some studies simply because it impacts negatively higher education (Zopiatis et al., 2015). Professional researchers leave their classrooms and course to low-ranked scholars affecting seriously the quality of the learning process. Furthermore, the new graduates enter a vicious circle given their lack of experience in professional research (Korstanje, 2016, 2022). Some of the problems professional researchers come through to abandon the teaching process are enumerated just below.

The vocational perspective of researchers is simply distorted by specific or externally-designed demands. To publish in top-tiered journals is a task very hard to perform even for consolidated scholars. There are several barriers and constraints in the act of publishing. In turn, leading journals keep a higher rejection rate based on more than 95%. The pressures for publishing make scholars to change regularly their objects of study. In normal conditions, teachers and researchers set their agenda revolving around their themes of interest. The doctrine of publishing or perish obliges authors to change their theme of research to be highly cited. In other cases, leading journals organize special issues to set an agenda in scholarship. In consequence, researchers are pressed to investigate topics that remain unattractive for them. Secondly, this lack of interest in the researched topic is chiefly associated with the time to teach lead professors to abandon teaching as well as interaction with students. Third and most importantly, there was a gap between professionally researching and teaching which is unfulfilled. The knowledge production published in leading journals never impacts the student's curriculum. This occurs because the paths of professional researchers and educators never cross again. Working on an ivory tower, researchers are rushed to publish their obtained outcome –on demand-whilst teachers follow an outdated curriculum. For that, the quality of higher tourism education ultimately declines because of the new global risks and challenges that place the industry in jeopardy (Korstanje, 2022; Korstanje & George, 2022). In the next part, we describe how the flow experience may very well lead to well-being and a state of joy.

3.5 From the Flow Experience to Joy

Interesting studies have emphasised on the outcome obtained by flow experience and positive psychology over the recent years. Mihali Csikszentmihalyi (1988) has innovated on the concept of flow experience. Per his stance, the concept of flow successfully activates the conditions towards an optimal experience. One might speculate that flow not only means a state of conscience but also in how a person may absorb knowledge. Ultimately, pleasure can be reached in the moment the feeling of transcendence can be controlled. Starting from the premise that happiness is not an objective condition of the self, he argues eloquently that it is a form of interpretation the surrounding environment. Those subjects who can control the emotionality of inner-experiences accepting their fears and negative emotions have further probabilities to reach happiness than those who are unfamiliar with the inner-world. At the same time, those persons who reach happiness are open to new experiences which lead them to potentiate their skills and cognitive features. In this vein, enjoyment occupies a central position in the configuration of a sustainable higher education curricula. Here, we have to discuss the essence of pleasure. Cskszentmihalyi acknowledges that the best experience is not completely pleasurable at a first instance, there are underling sentiments such as anxiety, fear or ambiguity which should be monitored. The concept of *optimal experience* corresponds with the phycological energy is divested to achieve certain real goals matching skills with opportunities for action; furthermore, the process opens the doors to more resilient personalities which are not only adapted to social change but also employ creativity to solve specific-based problems (Csikszentmihalyi, 1988; Csikszentmihalyi et al., 2014; Nakamura & Csikszentmihalyi, 2014). The concept of flow can be successfully applied to improve the professor-students' relations. In this token, some interesting studies have approached the gamification and development games such as FLIGBY (Buzady & Almeida, 2019; Gopher et al., 1994). Gamification theory offers a fertile ground to be widely applied to classrooms and the learning process. FLYGBY is employed as a serious game which exploits the individual skills and capacities of adaptation with entrepreneurship. This particularly has direct benefits for students and workers mainly involved in business environments. Serious games encourage students whereas alleviating their daily burdens. In fact, entrepreneurship optimises the decision-making process stimulating students' engagement in the learning process. The amassed knowledge should include creativity, critical thinking, leadership, negotiation and problem-solving capacities. To some extent, creativity is not enough to learn students the impacts of the actions (ethics). In this case, serious games enter in scene enhancing psychological motivation whilst adopting an ethical perspective. Serious games stimulate players' motivations recreating the background for emulating states of self-achievement and joyness. These skills call to perform an activity mainly marked by a certain degree of difficulty which causes positive and negative feedback in players. FLIGBY paves the pathway for the rise of a combined balance between harmony and self-achievement (Almeida & Buzády, 2019). In the same direction, interesting comparison and inferences can be done between FLYGBY AND PANCOE to overcome the barriers of higher tourism education. Unfortunately, application of FLYGBY in the fields of tourism and hospitality remain uncover or marginally explored (Almeida et al., 2021; Xu et al., 2016). Henceforth, it is important to describe PANCOE to fill the gap and getting applicable insights that improve tourism education in the next years.

3.6 Research Agenda for Tourism Education

It is noteworthy that tourism education has grown and evolved in basis of an economic-centred paradigm. From its outset, marketing and management took the lead in the configuration of the first tourism and hospitality-related careers. In the threshold of time different disciplines and subdisciplines have turned the attention to tourism as their main object of study. Social sciences such as psychology, anthropology, geography and sociology have achieved considerable steps in the study of tourism. As John Tribe (2010) laments, the maturation of tourism research has been cemented on a knowledge dispersion resulting from the lack of interest of the Academia to set a shared epistemological agenda. The collision of social science and the economic-centred paradigm remains open to date. Whilst the former sees tourism as a social institution which traverses cultures the later discusses tourism embracing enthusiastically the economic perspective. For these scholars, tourism should be defined as an industry or subservice sector maturated just after the WWII end. An additional tradition known as the Critical Turn has approached to the tourism research over the recent years. Furthermore, the pressure to publish recently imposed by educational institutions displace professional researchers from classic teaching. One of the troublesome aspects that led the higher tourism education to a state of crisis was the knowledge fragmentation resulted from the lack of consensus amongst the traditional academic schools. There is a type of gap between theory and practice which has not been filled out. PANCOE as well as other alternative techniques like FLYGBY provides with a robust framework to expand the current understanding of tourism education whereas facilitating answers to the current crisis.

3.7 Conclusion

As explained in former chapters, modern education appears to experience a serious crisis that affected curricula formation. Needless to say, tourism higher education is not an exception. From the outset, tourism education –as well as the first epistemologists- worked hard to lay the foundational tenets of tourism research. In the threshold of time, some critical studies have alerted on what J. Tribe named as "the indiscipline of tourism" which reflected a lack of order in knowledge production. This type of indiscipline was mainly marked by the monopolization of marketing and management pushing other disciplines to the margins. Although theories in tourism education have proliferated over the recent decades, we offer a conceptual framework gathering some of them in three clear-cut traditions: a) the epistemological tradition, b) the economic-centred tradition, and c) the critical tradition. From different angles, each one provided a firm background to understand tourism education. The epistemological tradition starts from the premise that researchers should set a palpable object of study. In so doing, the etymological roots of tourism should be found. Even if they did not reach any consensus on the nature of tourism, the epistemological tradition lays the basis to cement professional tourism research as well as for higher education. Rather, the economic-centred family concentrates efforts in setting agenda in the curricula formation. Instead of studying the social roots of tourism, this tradition explores new paths to adapt the students' curriculum to the real problems of the sector. More interested in protecting the interests of the industry –than understanding tourism as a social issue, the economic-centred theory has taken the lead in designing not only the curricula but also careers, syllabuses and a whole portion of the educational content. This moot point led the critical tradition to exert a caustic critique alerting on the limitations of considering tourism as a business only. The critical tradition lamented the hegemonic position of the "the managerial perspective" warning on the dissociation between professional research and teaching. For the sake of clarity, professional researchers are pressed to publish their advances leaving the classroom hours to less experienced scholars. The introduction of metrics, which substantially changes professional research, as well as the pressures for publishing in top-tiered journals, has created an irreversible chasm between teaching and researching. Gamification theory gives an interesting background to improve the student-professor relations, a problem which remains uncovered in the specialised literature. Henceforth, it is vital to discuss to what extent this gap affects the professor-student relationships in the fields of tourism (a point to be developed in the epilogue or last section). Ultimately, we have introduced some preliminary ideas on the concepts of flow, gamification theory and some commonalities between FLIGBY and PANCOE.

References

Airey, D. (2016). Tourism education: Past, present and future. *The Business of Tourism, 17*, 9–12.

Airey, D., & Tribe, J. (Eds.). (2006). *An international handbook of tourism education*. Abingdon.

Airey, D., Tribe, J., Benckendorff, P., & Xiao, H. (2015). The managerial gaze: The long tail of tourism education and research. *Journal of Travel Research, 54*(2), 139–151.

Akoglu, T. (2015). Walter Hunziker–the founder of academic tourism studies. *Anatolia, 26*(3), 501–505.

Almeida, F., & Buzády, Z. (2019). Learning entrepreneurship in higher education through flow theory and FLIGBY game. *International Journal of Virtual and Personal Learning Environments (IJVPLE), 9*(1), 1–15.

Almeida, F., Buzady, Z., & Ferro, A. (2021). Exploring the role of a serious game in developing competencies in higher tourism education. *Journal of Hospitality, Leisure, Sport & Tourism Education, 29*, 100347.

Amoah, V. A., & Baum, T. (1997). Tourism education: Policy versus practice. *International Journal of Contemporary Hospitality Management, 9*(1), 5–12.

Ateljevic, I., Harris, C., Wilson, E., & Collins, F. L. (2005). Getting 'entangled': Reflexivity and the 'critical turn'in tourism studies. *Tourism Recreation Research, 30*(2), 9–21.

Ateljevic, I., Morgan, N., & Pritchard, A. (Eds.). (2013). *The critical turn in tourism studies: Creating an academy of hope* (Vol. 22).

Baum, T. (2011). Global tourism higher education: The British Isles experience. In C. Hsu (Ed.), *Global tourism higher education: Past, present and future* (pp. 27–38). Abingdon.

Beckendorff, P., & Zehrer, A. (2017). *Handbook of teaching and learning in tourism*. Edward Elgar Publishing.

Belhassen, Y., & Caton, K. (2009). Advancing understandings: A linguistic approach to tourism epistemology. *Annals of Tourism Research, 36*(2), 335–352.

Benckendorff, P. (2009). Themes and trends in Australian and New Zealand tourism research: A social network analysis of citations in two leading journals (1994–2007). *Journal of Hospitality and Tourism Management, 16*(1), 1–15.

Boley, B. B. (2011). Sustainability in hospitality and tourism education: Towards an integrated curriculum. *Journal of Hospitality & Tourism Education, 23*(4), 22–31.

Botterill, D. (2001). The epistemology of a set of tourism studies. *Leisure Studies, 20*(3), 199–214.

Buzady, Z., & Almeida, F. (2019). FLIGBY—A serious game tool to enhance motivation and competencies in entrepreneurship. *Informatics, 6*(3), 27–35.

Carr, N., & Hayes, S. (2017). A comparison of tourism PhD Students' publication records and university of study. *Tourism Management Perspectives, 23*, 151–153.

Caton, K. (2014). Underdisciplinarity: Where are the humanities in tourism education? *Journal of Hospitality, Leisure, Sport & Tourism Education, 15*, 24–33.

Chon, K., Park, E., & Zoltan, J. (2020). The Asian paradigm in hospitality and tourism. *Journal of Hospitality & Tourism Research, 44*(8), 1183–1202.

Coles, T. (2009). Tourism studies and the governance of higher education in the United Kingdom. *Tourism Geographies, 11*(1), 23–42.

Coles, T., Hall, C. M., & Duval, D. T. (2006). Tourism and post-disciplinary enquiry. *Current Issues in Tourism, 9*(4–5), 293–319.

Coles, T., Hall, C. M., & Duval, D. T. (2009). Post-disciplinary tourism. In J. Tribe (Ed.), *Philosophical issues in tourism* (pp. 80–100). Channel View.

Cooper, C. (2002). Curriculum planning for tourism education: From theory to practice. *Journal of Teaching in Travel & Tourism, 2*(1), 19–39.

Cooper, C., & Shepherd, R. (1997). The relationship between tourism education and the tourism industry: Implications for tourism education. *Tourism Recreation Research, 22*(1), 34–47.

Costa, C., Carvalho, I., Cacador, S., & Breda, Z. (2016). Future higher education in tourism studies and the labor market: Gender perspective on expectations and experiences. In D. Prebežac, C. Schott, & P. Sheldon (Eds.), *The tourism education futures initiative: Activating change in tourism education* (pp. 193–213). Routledge.

Csikszentmihalyi, M. (1988). The flow experience and its significance for human psychology. *Optimal experience: Psychological studies of flow in consciousness, 2*, 15–35.

Csikszentmihalyi, M., Abuhamdeh, S., & Nakamura, J. (2014). Flow. In *Flow and the foundations of positive psychology* (pp. 227–238). Springer.

Cunliffe, S. (2002). Forecasting risks in the tourism industry using the Delphi technique. *Tourism, 50*(1), 31–41.

Dredge, D. (2015). Tourism and governance. In *Education for sustainability in tourism* (pp. 75–90). Springer.

Du, J. (2003). Reforms and development of higher tourism education in China. *Journal of Teaching in Travel & Tourism, 3*(1), 103–113.

Dwyer, L. (2012). Trends underpinning global tourism in the coming decade. In W. Theobald (Ed.), *Global tourism* (pp. 542–558). Routledge.

Ertaş, M., & Kozak, M. (2020). Publish or perish: The proportion of articles versus additional sections in tourism and hospitality journals. *Journal of Hospitality and Tourism Management, 43*, 149–156.

Fernandez Fuster, L. (1967). *Teoría y Técnica del turismo. [Theory and practice of tourism] Tomo I*. Editorial Nacional.

Fidgeon, P. R. (2010). Tourism education and curriculum design: A time for consolidation and review? *Tourism Management, 31*(6), 699–723.

Gopher, D., Well, M., & Bareket, T. (1994). Transfer of skill from a computer game trainer to flight. *Human Factors, 36*(3), 387–405.

Gretzel, U., & Bowser, G. (2016). Real stories about real women: Communicating role models for female tourism students. In D. Prebežac, C. Schott, & P. Sheldon (Eds.), *The tourism education futures initiative: Activating change in tourism education* (pp. 214–227). Routledge.

Gursoy, D., & Sandstrom, J. K. (2016). An updated ranking of hospitality and tourism journals. *Journal of Hospitality & Tourism Research, 40*(1), 3–18.

Hall, C. M. (2011). Publish and perish? Bibliometric analysis, journal ranking and the assessment of research quality in tourism. *Tourism Management, 32*(1), 16–27.

Hall, T., Gray, T., Downey, G., Sheringham, C., Jones, B., Power, A., & Truong, S. (2016). Jafari and transformation: A model to enhance short-term overseas study tours. *Frontiers: The Interdisciplinary Journal of Study Abroad, 27*, 33–46.

Houlot, A. (1961). Le Turisme et La Biblie. Revue l'Académie Internationale du Turisme. Monaco.

Hsu, C. H. (2018). Tourism education on and beyond the horizon. *Tourism Management Perspectives, 25*, 181–183.

Inui, Y., Wheeler, D., & Lankford, S. (2006). Rethinking tourism education: What should schools teach. *Journal of Hospitality, Leisure, Sport and Tourism Education, 5*(2), 25–35.

Jafari, J. (2001). The scientification of tourism. In V. L. Smith & M. Brent (Eds.), *Hosts and guests revisited: Tourism issues of the 21st century* (pp. 28–41). New York.

Jafari, J. (2002). *Tourism education and training models-getting to the core of destination planning and management* (pp. 29–34). World Tourism Organization.

Jafari, J., & Ritchie, J. B. (1981). Toward a framework for tourism education: Problems and prospects. *Annals of Tourism Research, 8*(1), 13–34.

King, B., McKercher, B., & Waryszak, R. (2003). A comparative study of hospitality and tourism graduates in Australia and Hong Kong. *International Journal of Tourism Research, 5*(6), 409–420.

Korstanje, M. E. (2007). The origin and meaning of tourism: Etymological study. *E-Review of Tourism Research, 5*(5), 100–108.

Korstanje, M. E. (2015). A portrait of Jost Krippendorf. *Anatolia, 26*(1), 158–164.

Korstanje, M. E. (2016). Why citation impact is so important for tourism researchers? *Advances in Hospitality and Tourism Research (AHTR), 4*(1), 26–31.

Korstanje, M. E. (2022). Tourism imagination: A new epistemological debate. *Current Issues in Tourism*, 1–13. Ahead of print. https://doi.org/10.1080/13683500.2021.2023481

Korstanje, M. E., & George, B. (2022). *The nature and future of tourism: A post-COVID-19 context*. Apple Academic Press.

Krippendorf, J. (2010). *Holiday makers*. Routledge.

Ladkin, A. (2014). Employment and career development in tourism and hospitality education. In D. Dredge, D. Airey, & M. Gross (Eds.), *The Routledge handbook of tourism and hospitality education* (pp. 427–439). Routledge.

Lam, T., & Xiao, H. (2000). Challenges and constraints of hospitality and tourism education in China. *International Journal of Contemporary Hospitality Management., 12*(5), 291–295.

Law, R. (2017). Ranking individuals in tourism and hospitality research. *Tourism Recreation Research, 42*(3), 392–397.

Leiper, N. (1979). The framework of tourism: Towards a definition of tourism, tourist, and the tourist industry. *Annals of Tourism Research, 6*(4), 390–407.

Leiper, N. (1983). An etymology of "tourism". *Annals of Tourism Research, 10*, 277–281.

MacLaurin, D. (2006). Tourism education in Canada: Past, present and future directions. In C. Hsu (Ed.), *Global tourism higher education: Past, present and future* (pp. 1–26). Abingdon.

McKercher, B. (2002). The future of tourism education: An Australian scenario? *Tourism and Hospitality Research, 3*(3), 199–210.

McKercher, B. (2008). A citation analysis of tourism scholars. *Tourism Management, 29*(6), 1226–1232.

Moscardo, G. (2011). The role of knowledge in good governance for tourism. In E. Laws, H. Richins, & J. Agrusa (Eds.), *Tourist destination governance: Practice, theory and issues* (pp. 67–80). CABI.

Nakamura, J., & Csikszentmihalyi, M. (2014). The concept of flow. In *Flow and the foundations of positive psychology* (pp. 239–263). Springer.

Paris, C. M. (2011). Social constructivism and tourism education. *Journal of Hospitality, Leisure, Sports and Tourism Education (Pre-2012), 10*(2), 103–115.

Pearce, P. (2011). Australian tourism education: The quest for status. In C. Hsu (Ed.), *Global tourism higher education: Past, present and future* (pp. 251–268). Abingdon.

Pechlaner, H., Zehrer, A., Matzler, K., & Abfalter, D. (2004). A ranking of international tourism and hospitality journals. *Journal of Travel Research, 42*(4), 328–332.

Prebežac, D., Schott, C., & Sheldon, P. (Eds.). (2016). *The tourism education futures initiative: Activating change in tourism education*. Abingdon.

Rakić, T., & Chambers, D. (Eds.). (2012). *An introduction to visual research methods in tourism*. Routledge.

Reichel, A. (2011). Tourism and hospitality higher education in Israel. In C. Hsu (Ed.), *Global tourism higher education: Past, present and future* (pp. 61–88). Abingdon.

Ryan, C. (2005). The ranking and rating of academics and journals in tourism research. *Tourism Management, 26*(5), 657–662.

Sheldon, P. J., Fesenmaier, D. R., & Tribe, J. (2011). The tourism education futures initiative (TEFI): Activating change in tourism education. *Journal of Teaching in Travel & Tourism, 11*(1), 2–23.

Stevens, Z. M., Grimwood, B. S., & Caton, K. (2019). Story and moral development in tourism education. *Journal of Teaching in Travel & Tourism, 19*(1), 22–38.

Tribe, J. (1997). The indiscipline of tourism. *Annals of Tourism Research, 24*(3), 638–657.

Tribe, J. (2000). Indisciplined and unsubstantiated. *Annals of Tourism Research, 27*(3), 809–813.

Tribe, J. (2001). Research paradigms and the tourism curriculum. *Journal of Travel Research, 39*(4), 442–448.

Tribe, J. (2004). The indiscipline of tourism. In S. Williams (Ed.), *Tourism: Critical concepts in the social sciences, 2* (pp. 205–226). Routledge.

Tribe, J. (2008). Tourism: A critical business. *Journal of Travel Research, 46*, 245–255.

Tribe, J. (2010). Tribes, territories and networks in the tourism academy. *Annals of Tourism Research, 37*(1), 7–33.

Wang, J., Ayres, H., & Huyton, J. (2010). Is tourism education meeting the needs of the tourism industry? An Australian case study. *Journal of Hospitality & Tourism Education, 22*(1), 8–14.

Weber, K., & Ladkin, A. (2008). Career advancement for tourism and hospitality academics: Publish, network, study, and plan. *Journal of Hospitality & Tourism Research, 32*(4), 448–466.

Xiao, H. (2013). Jafar Jafari: The platform builder. *Anatolia, 24*(2), 288–296.

Xu, F., Tian, F., Buhalis, D., Weber, J., & Zhang, H. (2016). Tourists as mobile gamers: Gamification for tourism marketing. *Journal of Travel & Tourism Marketing, 33*(8), 1124–1142.

Yankholmes, A. K. (2014). Publish or perish: African scholarship in the field of tourism and hospitality studies. *Tourism and Hospitality Research, 14*(1–2), 97–107.

Young, T., & Maguire, A. (2017). Indigenization of curricula: Trends and issues in tourism education. In P. Beckendorff & A. Zehrer (Eds.), *Handbook of teaching and learning in tourism*. Edward Elgar Publishing.

Zhao, W., & Ritchie, J. B. (2007). An investigation of academic leadership in tourism research: 1985–2004. *Tourism Management, 28*(2), 476–490.

Zopiatis, A., Theocharous, A. L., & Constanti, P. (2015). 'The past is prologue to the future': An introspective view of hospitality and tourism research. *Scientometrics, 102*(2), 1731–1753.

Chapter 4
Marketing Education in Times of COVID-19: Argentina as Main Study Case

Abstract Doubtless, COVID-19 has affected not only the tourism industry as well as global trade but also education worldwide. Although technology occupied a central position in online classroom, no less true was that many students had not access to digital platform. This chapter interrogates on the problem of curricula in higher tourism education for Argentina to understand the new global risks. The pandemic has opened the doorsteps to new opportunities, problems and challenges educators should bear on mind. In this vein, creativity and entrepreneurship play a leading role in the formation of a new curricula in a post COVID-19 context. To think the future research lines of tourism and hospitality in a post-COVID-19 context is very hard to grasp, simply because the future of tourism seems to be a bit uncertain. In fact, in this post-COVID-19 world, new technologies lead us to a landscape without tourists. At the same time, digital technology helps in the creation of more resilient and smarter destinations.

Keywords Resiliency · Global risk · COVID-19 · Tourism education · Adaptancy

4.1 Introduction

Although tourism marketing and management have advanced considerable firm steps recently the crisis precipitated by COVID19 has posed serious challenges and risks for the sector, above all for the education in tourism fields. The number of scholars, who launched to study the effects of COVID19 in the tourism industry, has notably increased (Prayag, 2020). Despite the books, studies, and working papers revolving around this theme, theories may be grouped into two clear-cut families. Those studies interested to see the COVID19 as a great opportunity for the rebirth that alleviates the consequences of climate change –if not the ecological crisis (Gössling et al., 2020; Higgins-Desbiolles, 2020a, 2020b; Nepal, 2020; Sigala, 2020), and those moved to forge new methodologies and pedagogic instrument at universities, so to speak, in a world without tourists (Hayes, 2020; Korstanje & George, 2020; Wen et al., 2020). Whilst the imposed lockdowns, as well as the

A. Zuccoli, M. E. Korstanje, *The Role of Pleasure to Improve Tourism Education*, https://doi.org/10.1007/978-3-031-21580-3_4

restrictive measures around global trade and sub-service sectors, had harmed the tourism industry worldwide, no less true is that the access to digital media, ICT and new emerging technology is revolutionizing education in the times of pandemic as never before (Gretzel et al., 2020). Zeng et al. (2020) recognise that artificial intelligence, technology and robots have gained incredible tractions since the adoption of the restrictive travel bans that affected the tourism industry because of COVID-19 (Omar Parvez & Cobanoglu, 2021). Robots and digital technology not only reduce human contact but also helps to sterilize public spaces. However, some controversies sparked according to the belief the rise of technology results in the losses of jobs for the industry. History witnesses, as authors assess, how the innovations of new technologies are applied to the advance of humanity. Quite aside from any controversy, robots and technology will change the tourism industry in terms of expectations, service quality and image (Lew et al., 2020). Digital technology offers a fertile playground to change the current programs, courses and syllabuses in tourism and hospitality rechannelling them into a new more sustainable direction. Authors, who support this idea, signal to the role of resiliency in creating safer and more adaptable destinations to the devastating effects of COVID-19 in the years to come (Carr, 2020; Haywood, 2020). Beyond any discrepancy, these studies apply for the Global North and little evidence exists on how the "Global South" is prepared for performing such a task.

In this complex context, the present study-case explores the existing incongruence in professional tourism research as well as the importance of bridging the gap between theory and practices in the field, above all the problems of employability in the tourism and hospitality industries in a post-COVID-19 context. Needless to say that COVID-19 jolted the West from its lingering complacency evincing the material inequalities of global capitalism. The current chapter is based on Argentina as main study case. Education not only occupies a central position in the training of future tourism professional but also situates as the touchstone of the discipline for the hard times to come. The goals of this study-case are threefold. We dissect the complex nature of tourism education, as well as the current limitations in tourism research. Additionally, we lay the foundations towards a new understanding of the opportunities beyond the COVID19 crisis. Lastly, an illustrative study-case bridging the gap between theory and practice is provided.

4.2 Tourism Education

As it was debated in the earlier sections of this book, the term education stems etymologically from the Lat word *educere* which was ambiguously used not only to denote *emancipation* but also to feed the cattle. This suggests the meaning of education rests on two contrasting forces. On one hand, it helps subjects in going forwards against the status quo. On another, it domesticates students within the frames of established paradigms (Bass & Good, 2004). In the tourism industry education has taken a managerial perspective developing what Airey et al. dubbed as *managerial gaze* (Airey et al., 2015). Such a position coincides with the long-dominant paradigm at higher education and many universities which educate professional for getting a job in the industry instead of scientists or researchers (Jennings, 2001; Paris, 2015; Xiao & Smith, 2010). This creates, of course, a gap between a theory which is often discouraged by policy-makers given its lack of pragmatism, and practice. Since the theory seems to be what Academicians look for, tourism practice is valorised by some voices as a vehicle towards more sustainable and resilient destinations (Costa, 2001; Ruhanen, 2006; Wang et al., 2010). In this respect, John Tribe toys with the belief that the tourism research is facing a type of dispersion, he dubbed as the *indiscipline of tourism* denoting certain lack of coherence of the dominant academic theories in applied research. To put this simply, the evolution of tourism research coincides with clear anarchy in the knowledge formation process which obscures more than it clarifies. Instead of being dominated by a guiding paradigm, the discipline is centred on countless approaches and academic traditions which forge grounded theory beyond the intervention of the Academy (Tribe, 1997, 2000, 2010; Tribe & Liburd, 2016). To the problems of dialogue between the Academy and the industry, we must add the limitations of universities to meet the labour market demands. The figure of employability plays a leading role by training and educating the future professionals of the sector (Inui et al., 2006). What seems to be more important, the negative –and durable- effects of COVID19 in the industry are re-shaping tourism education as never before. Qiu et al. (2020) explore the opportunities given by technology to facilitate social distancing at schools and universities. Doubtless, one of the problems or dilemmas of tourism education lies in the lack of certainness or knowledge about what would be the future jobs the industry will demand. Not surprisingly, the dynamism of tourism industry is changing every day and of course just after the turn of the twentieth century (Benckendorff et al., 2010). Christou and Sigala (2001) recognise that it is very hard to precise an accurate research agenda for tourism education simply because the nature of the service sectors. With focus on tourism education, they argue convincingly that online education offers pros and cons regarding practical reference and the revisions of specific study-cases. Having said this, Higgins-Desbiolles (2020a, 2020b) acknowledges that the pandemic has been a game-changer for the industry, presenting an opportunity for a binary debate between a rebirth –or re-foundation- and recovery. The tourism industry has benefits but at the same time costs, as Higgins-Desbiolles adheres. It is safe to say that the pandemic leads scholars to think new methodologies and instruments to overcome the current methodological setbacks. Wen et al. (2020) alert that it is the time to adopt multidisciplinary methods and approach to face the challenges pose by the pandemic. Stepping back to the same line of inquiry, Tiwari et al. (2020) stress many studies focus on the economic impact of COVID19 in the subservice sector but little attention is paid to the devastating consequences in tourism education. Furthermore, there is a chasm between academics and industry which merits to be filled. In the same way, the employers' opinions dissociate often from the hopes and premises of recruited students. Per authors' argumentation, COVID19 ignites a hot debate revolving around opportunities of new learning and the foundation of new vocational skills to meet the requirements of the industry. The curriculum designs should be balanced for students to deal with theoretical questions as well as designing programs to assist industry stakeholders. In the same direction, Ratten and Jones (2020) overtly say that the future of tourism is given by the entrepreneurship education and the efficiency of emerging practice paths. In an ever-changing society, entrepreneurial education guides policy-makers to cope with new solutions for the industry that a post covid19 context finally requests. Scholars misjudge the nature of recovery and change. Whilst the former signals to stable conditions of growth just after a disaster takes hit, –probably keeping the foundational cultural values- the latter refers to a new radical opportunity for the change. As they eloquently observe, the crisis opens the doors to response mechanisms which alternate the change with stability.

4.3 Tourism, Technology and Resiliency

The figure of technology has a great controversy in the constellations of social science and pedagogy. Some studies allude to technology as an efficient instrument to create smarter destinations and more resilient enterprises (Jeong & Shin, 2020; Xiang et al., 2015). Others, rather, exert a caustic critique on the role of technology generating automated behaviour in consumers (Korstanje & Seraphin, 2018). Technology may be very well used to avoid disaster or mitigate its negative impacts in many ways. For some reason, in the fields of tourism technology seems not to be employed to make more resilient destinations. Technologies facilitate the process of information management to make the most efficient decision in minutes (Jacobsen & Munar, 2012). Digital technologies offer a fertile ground towards the development of safer and more resilient destinations. Having said this, the access to technology seems to be far from being homogenised in the global North and South. Under some conditions, technology aggravates the already-existent material asymmetries of society. What is equally important, the workforce as well as the working union may seen technology with certain suspect. There are interesting points of convergence on the mistrust caused by the adoption of robots as housekeepers at hotels (Ivanov & Webster, 2020); in a nutshell the debate is far from being closed. In a seminal book which entitles *Laws of Fear* Cass Sunstein calls the attention on the complex relation between risk perception and technology. Starting from the premise that the human mind is subject to emotional distortions which often led us to misjudgements about risks, Sunstein argues convincingly that technology potentiates the human character helping in the planning of post disaster context, but emotionality often leads us to over-valorise some risks whilst others major ones are overlooked. He put the example of different experiments where lay-people was invited to emulate a decision-making process in context of disaster. As Sunstein eloquently notes, lay-people made bad decisions in the experiment in so far as the net of experts. This means that no matter the qualification and education, technology is not enough to overcome the cognitive distortions (Sunstein, 2005). However, technology still remains as a promising tool to be used in post disaster contexts.

As the previous argument is given, Gretzel and Scarpino-Johns (2017) hold digital technologies not only helps policy makers to predict future disasters but also contributes notably to the creation of smart tourism destinations. Likewise, smart tourism increases competitiveness whilst supporting sustainability goals. For authors, the concept of resiliency and sustainability are inextricably intertwined. Therefore, as they lament, scholars have not paid enough attention to destination resilience. Smarter destinations build and equip communities with sensing, opening, sharing, governing and innovating capacities to improve destination resilience. According to this, Amore et al. (2018) punctuate on the different levels of resilience, an idea that escapes to the presence of a single accepted definition of the term. Resilience should be constructed adopting multilevel models which evince very well the complexity of multiple and complex factors interacting in the different levels of the system. The first level consists in a *macro-landscape development* which means the articulation of technological and socio-cultural factors including technological changes, population and politics. Resilience cannot be successfully planned without taking into account the recent digital revolutions worldwide. The second level is *the meso-sociotechnical landscape.* It suggests the interaction of multifactor network which combines explicit rule with implicit customs or habits. The third level understands those creative actors who give innovative behaviour to the industry (authors cite the example of peer-to-peer accommodation providers as Couchsurfing or Airbnb). These authors start from the premise any state of disaster opens the doors to innovative opportunities to optimize the destination performance towards a resilient industry. In a hyper-mobile world, digital technologies –through the lens of governance- offer more predictable landscape to optimize the decision-making process. At the same time, political leadership plays a leading role legitimating resilient strategies. In a seminal book which resulted as inspiration after the 2010–2011 earth-quake that whipped Christchurch, New Zealand, Hall et al. (2017) contend that there is an inevitable tension between the academic theory and the practical reactions of community members –or in their terms the on-the-ground experience-. The question of resilience stabilises the complex interaction (discrepancies) among the different levels and sub-systems which include travel networks, organizations, locals, business and so forth. Technologies innovate in one direction creating new opportunities whilst preserving habits and stability in another direction. But things come worse to worst precisely because new risks and problems can emerge. As Hall, Prayag & Amore infer, the fourth industrial revolution, which mainly marked a global economic growth, as well as a rapid automation of production, ensures a global integration among nations, but at the same time, it gives some problems to the domestic labour or the economic market deregulation. With robots at the factories, the future of workers remains uncertain. Ivanov & Webster alert on the advantages and disadvantages of robots employed at tourist destinations and hotels; the *robonomic economy* which denotes the massive introduction of artificial intelligence and robots to serve guests, denote a much deeper process of automation technology re-optimizing the economic process of production and consumption. To some extent, these radical changes give stability to the system. Unfortunately, it creates profound changes in the levels of politics and social relations (Ivanov &

Webster, 2019; Webster & Ivanov, 2020). Lastly, some studies have emphasised on the importance of ethics in guiding policy-makers to promote positive changes for locals. Disasters often place the industry in a state of crisis which unless dully resolved may lead the destination to a decline facet. Resilience applied to post-tourism crisis situates as a necessary topic which populates in an increasing number of syllabuses and tourism-related courses across the globe (Séraphin et al., 2019; Walters et al., 2016). This raises a more than interesting question: what is and how the term resilience has evolved in the tourism literature?

4.4 The Power of Resiliency in Post Tourism Crisis

The term resiliency was originally coined by Victor Frankl, a Nazi concentration camp survivor, whose studies were oriented to study how a person overcomes the diverse obstacles and setbacks in the lifetime. Resiliency means the abilities of somebody to adapt to hostile or adverse situations. Psychology has historically referred to the term as an ability of invoked by subjects to cope with traumatic experiences. Over the recent years, disaster management sheds light on the importance of resilience in the recovery facet –just after a disaster takes hit- (Amaratunga & Haigh, 2011; Kapucu & Khosa, 2013). In the tourism fields, resilience mainly associates with the capacity of tourist destinations to cope with and overcome different crises provoked by external or internal threats. Just after the turn of the century, countless works emphasised the needs of creating more resilient destinations (Cochrane, 2010; Korstanje & Ivanov, 2012; Light, 2017; Tyrrell & Johnston, 2008). Whilst risk perception theory focused on the urgency of locating and eradicating future risks, post-disaster tourism –as an emergency discipline- delved into the importance to adapt to risky situations. Adaptation, not prevention, was the new buzzword that invaded the top-ranked journals worldwide (above all, in English speaking countries) (Seraphin et al., 2020). This new surfacing sub-discipline alludes to the opportunity to learn a lesson of the causes that led to disaster by the application of different methods and instrument in so far accelerating the recovery timeframe through the stimulation of tourism consumption. Thousands of new visitors launch to visit spaces devastated by disasters, traumatic events or in an underway reconstruction. Examples like New Orleans, Sri Lanka, Chile, Japan, and Haiti (only to name a few) and other sites probe how tourism acts as an instrument of resiliency in post-disaster contexts (Chew & Jahari, 2014; Gotham, 2017; Lin et al., 2018; Robbie, 2008; Tucker et al., 2017). Nonetheless of the fact post-disaster management occupies a central position in the agenda of future professional researchers, no less true some voices have alerted some limitations of the theory –at least as it is being formulated-. Post-disaster tourism has invariably shown interesting results helping communities in the post-recovery stage but some cases, giving a biased diagnosis of the problem or a partial version of history. The re-memorisation process is often subject to some distortions when the political power intervenes to give its version of the event. In other cases, post-disaster consumption leads gradually to "a spectacle of disaster" which says little on the real causes of the event. When this happens the possibilities for the disaster to take room anew turns higher (Korstanje, 2016; Sather-Wagstaff, 2016; Tzanelli, 2016). In next section, we analyse the syllabuses and programs of the most important universities have tourism and hospitality as Bachelor career.

4.5 Methodological Discussion

In Argentina, the hospitality and tourism careers have more than 4 decades of training the future professional of the industry which includes Tourism Bachelors, policymakers, as well as tour-operators, Tour guides or hotel staff members. Unlike other regions or neighbouring countries, Argentina lacks of Ph. Doctorate careers in tourism and hospitality. Furthermore, in the syllabuses, a managerial imprint prevails. The present section reviews the syllabuses of main Argentinean universities that offer tourism and hospitality as professional careers. Argentina has more than 40 public and private universities where students earn their degree in tourism. We review the syllabuses of 9 selected universities to determine to what extent courses related to post disaster recovery or resiliency are found. Universities have been selected according to the number of students, prestige and their leading presence in QS ranking. The nature of careers ad courses varies on university and region, for that we have standardised the different courses in 10 clear-cut families: Sustainability, geography, marketing and Management, Law and Right studies, Humanities, History, Methodology, foreign languages and risk management, governability and post disaster studies. The outcome is complemented with formal and informal interviews conducted in the field. We consulted 20 professionals of the industry as deans, rectors, and well-versed researchers. The sample was drawn by 10 males and 10 females ranging from 25 to 65 years old. Based on the snowball method, each person suggested the next interviewee to get a hold of. Last but not least, it is important to mention that the present research pursues an exploratory approach. So, the subthemes discussed by participants do not match necessarily with the state of the art or the conceptual discussion in earlier sections. They were proposed initially by interviewees.

4.5.1 The Study-Case

Per a preliminary review, Table 4.1 shows courses associated to economy predominates with 20.9% of the total of course, seconded by Geography (17.9%), foreign languages (15.72%), Marketing and Management (13.2%), sustainability (6.11%), humanities (7.86%), history (6.55%), methodology and applied research (6.11%), and law-jurisprudence (5.24%). The risk management or post-disaster studies occupies a marginal part with scarcely 0.44% of the total courses estimated. What is

Table 4.1 Number of courses by University in Argentina

University	Sustainab.	Geog.	Mark.	Law	Econ.	Human.	Hist.	Method.	Lang.	Risk Man.
U. Palermo	1	7	5	2	5	2	1	1	4	0
U. San Martin	3	7	5	2	7	2	4	1	4	0
U. La Plata	1	3	2	1	4	0	2	1	4	0
U. del Sur	1	2	1	1	4	1	1	1	4	0
U. Misiones	0	5	3	0	5	2	1	1	3	0
U. Comahue	3	1	5	1	5	4	0	2	4	0
U. Lanus	2	6	5	2	5	2	3	2	5	1
U. Mar del Plata	1	6	3	2	4	4	1	3	4	0
U. of Quilmes	2	8	5	1	7	1	2	2	4	0
Sub-totals	**14**	**45**	**34**	**12**	**46**	**18**	**15**	**14**	**36**	**1**
Percentages	**6.11**	**17.90**	**13.97**	**5.24**	**20.09**	**7.86**	**6.55**	**6.11**	**15.72**	**0.44**

Source: self-elaboration, 2021

equally important there are not courses related to the introduction of digital technologies to make more resilient destinations. Neither artificial intelligence nor the role of robots to mitigate disasters is certainly found in the inter-syllabuses analysis. This trend evinces that courses in tourism Bachelor are heavily influenced by economic-based courses whilst only one establishment, the University of Lanus has a course specialised in risk management. At the time of detail, the same information by university, we note that some universities as National University of Comahue and University of Mar del Plata are prone to have courses based on social science whereas University of Palermo, University of Quilmes and University of San Martin tend to geography as an important course in their syllabuses. The foreign languages keep stable during the tourism and hospitality BA.

A qualitative view reveals that syllabuses are not designed to cover risk management or post-disaster studies. Although authorities recognize the importance of adopting these new paradigms in an ever-changing world, where the industry faces many global risks, no less true is that syllabuses were originally designed in the mid of 90s decade where the managerial paradigm, –echoing Tribe- dominates. Tourism education in Argentina does not contemplate new global risks as terrorism or climate change; in fact, the courses based on climate crisis or sustainability does not abound. Only 6.14% of the reviewed syllabuses contain courses associated with sustainability. The University of Comahue is the establishment that more courses offer whilst the University of Misiones lacks courses relating to sustainability. Economy and geography (above all, the geography of tourist circuits) mark the pace in tourism education. This information was contrasted with 20 interviews with deans, rectors and professional tourism researchers using the snowball method which means that each interviewee gives recommendations to the next person to consult. Because the snowball method is not statistically representative the results –though illustrative- cannot be extrapolated to other universes.

4.5.2 Risk Perception Theory

The dissociation of the new reality of the industry and the old-fashioned syllabuses is present in a whole portion of participants. Risk perception theory has played a leading role in the tourism research since the turn of the century. The 9/11 associated to new global risks has captivated the attention of researchers. Risk perception theory explores the different facets and psychological conditions that lead consumers to target some destination by avoiding others. For some voices, 9/11 was a foundational event that multiplicated the risk perception studies worldwide. Although participants recognize the importance of risk perception studies in the elaboration of programs that mitigate the negative effects of the pandemic, they overtly lament universities in Argentina is overlooking the theme by a lack of interest. Matias (52 years old) is a professor at a well-known private University. Per his stance, "*after 9/11 the industry goes through many devastating events, risks or disasters. Future tourism policymakers should be trained in risk reduction programs or post-disaster recovery courses. Unfortunately, the syllabuses contents are far from being updated. This creates a gap between theory and practices in tourism, a gap very hard to fill out. I suggest my students read professional journals, above all journals published in English but students rarely do it. We lack an integral program of risk management for the tourism industry; the tourism ministry is not taking the lead path in this direction*". At a first glimpse, Ana, who is a professional researcher with a focus in tourism, coincides with this acknowledging: "*a couple of decades ago, when I was a student I never imagined the turn of the century would be characterised by frightening events. Would you figure out the world after COVID-19? Or how do we make good research in a world without tourists? If you ask me, this happens simply because the Academia remains back respecting to the urgencies of the industry, in a world in constant change. We train tour operators, or guides not professionals researchers. This is the reason why the economy and marketing-centered programs overweight social sciences (Ana 50 years old, female). Franco (28 years old) is a young researcher who claims the important of sociology over marketing courses, he goes on to say that "sociology is the basis of critical thinking. In tourism we need to create theory, produce a new understanding of the issue. Tourism is not only an industry, it is a social institution. Holidays are real rite of passage our parents and grand-parents do before us. The economic-based paradigm reduces the rich nature of tourism only to profits or what can be measured. When the industry is devastated, like the COVID-19 case, the paradigm fails to give new orientation to follow. This happens simply because they start from the premise tourism is only a industry. They echo this because they lack of formal knowledge in sociology or anthropology". Esteban (30 years old, Male) emphasizes that "humanism should be introduced and populated as powerful option in tourism careers. Humanism has the necessary ethical background to make more resilient destinations. Without ethics there is no tourism viable".*

4.5.3 (Digital) Technology

The interviewed participants mentioned the lack of courses or syllabuses content related to digital technologies. In this respect, they interrogate the opportunities given by the COVID-19 to employ new digital technologies in tourism education. Gradually the pandemic is pushing us to a virtual world where robots and artificial intelligence mediate to develop safer destinations. Here participants divide into two clear-cut groups. Those who signal to technology as a catalyst to boost tourism education and sustainability, and those who alert on the impossibility of technology and virtual tourism to replace the personal human experience which characterizes host guests encounters. For them, the tourism industry follows a relational nature which is indispensable for sustainable tourism. Pedro is a tour guide who works in Cordoba (Male 42 years old), he strongly believes technology helps but it is not determinant of tourism consumption, *"how do you think tour-guide will make their job? Do you believe robots are the solution? In this grim world the humanity will be replaced by machines and automats.. the tourism industry engages with human contact, the COVID-19 and particularly the social distancing will become our differences more acute. In this context, technology is not part of the solution but the problem. As a Tour guide, I not only show the attraction I get familiar with each personal history trying to understand why they (the visitors) are paying me attention when I guide. There is a good relationship between me and them which is now virtualised because of COVID19!*

Nonetheless of the fact that virtuality entails negative consequences for the industry, Esteban (CEO of an important hotel) replies: *"the use of technology is very fruitful for the industry, as well as for the short-run strategies. Not only technology allows safer destinations but also it contributes positively to improving tourism education. Less pollution is more sustainable tourism, in such a case, virtual tourism will help de-growth destinations making more sustainable spots. Climate change needs an urgent solution. We need more resilient, sustainable and smarter destinations. The present state of chaos and crisis gives us a perfect opportunity to change, the point is what changes to what direction?"* (Esteban, male, 49 years old). Unfortunately, he continues, tourism-related syllabuses never focus on the introduction of digital technologies to be applied in smart destinations. "*I receive everyday young people who love to work at my hotel. They are moved, motivated and engaged to the guest but I realize they have not been trained in big data, or digital platform to serve our guests. This is a big problem in the future tourism professional staff education!*

4.5.4 Resilience

Another (second) significant point of discussion in the consulted experts is the limitations of risk theory to prevent disasters. Jorge is dean of a national University. He recognizes that risk perception has been widely applauded in the Academia, above all in the leading journals but risk perception theory runs short at the time of explaining disasters. For Jorge, the disaster represents a unique and outstanding event where all responses of society simply fail. Hence, the notion of prevention which characterizes the conceptual core of the discipline should set the pace to *adaptation* which relates to resiliency. He overtly says: "*we need to design syllabuses according to the next world, a complex world where all diagnoses are incorrect. The risk theory alludes to a hypothetical scenario mathematically simulated. When the disaster takes hit, nobody follows the protocols. The COVID19 is a perfect example of what I want to say... neither authorities nor experts know what should be done to restart the tourism industry. For this, we need to educate students to construct more resilient destinations for the years to come*" (Jorge 59 years old, male). Maria is a well-known Dean and per her expertise, "*resilience probed to be useful in psychology, sociology and post disaster management. In the world, the tourism research advocated efforts to study this new subfield. In Argentina, it was unfortunate that programs lack of resilience-related courses that give students a fresh perspective on the next years. Figure out when I studied tourism all these global risks were watched in fiction movies. Now we are living in an ever-changing world which merits new reactions from our sides*". There is a clear passage from classic studies –for example when Jorge says "when I studied tourism", and the urgency of a new syllabus content which includes risk perception studies or disaster management.

4.5.5 Tourism Marketing and the Managerial Gaze

In third, some interviewees understand that marketing should not be demonised as the exegetes of social sciences claim. There is nothing like a discussion –tension-between the managerial perspective and social sciences. Though the latter signals critical reasoning which introduces benefiting changes in the system, it is vital to adopt a new marketing technique oriented to resiliency. Post-disaster marketing and post-disaster marketing education gravitate in the consulted professors, deans, and professional fieldworkers. As Ignacio puts it, "*the hot debate in favor or detriment of tourism management and social science does not work anymore. Now we need to change the paradigm thinking of new alternatives. Think in this twice ... marketing is not good, not bad, it is an instrument, the all-catch engineering piece that boosts or condemns destinations. We urge to educate leaders versed in post disaster marketing. In the same way, sociology opens the doors to understand the complexity of certain issues whilst Marketing –in dialogue with sociology- gives fresh insight to make more resilient destinations* (Ignacio, 62 years old, Male). For Christina, who serves as a tourism professor at a state university, there is a strange dichotomy between private and public education in tourism. Private education pioneered tourism education whilst state universities are almost new educating tourism professionals. The economic-centered paradigm is originated in private universities in the 80s decade whilst borrowed from the state universities in the 90s. State universities borrowed the syllabuses content from the private sector conserving a managerial perspective: *"though tourism sociology or anthropology are not being accepted as leading courses in the tourism fields, state universities keep more disciplines linked to social humanism. History of art or anthropology of tourism is not given to students of private universities. Private Universities educate students to be professionals, or policy makers whereas state Universities train professional fieldworkers. I feel that in both cases the problem of governability, risk management or resiliency is overlooked by the tourism ministry. Training tour guides was an option of the past when the tourism machine was based on sun and beach, but now new forms of tourism consumption emerge. I speak you of dark tourism, War tourism, Slum tourism or creative tourism. The industry is changing and new times need a new solution. We need to train well-versed experts based on the decision-making process in case of emergency or specialised in post-disaster recovery.. I am fed to hear of the history of Mar del Plata or Bariloche; I want to hear what would be done about COVID19"* (Christina, 33 years old, female). Roberto –professor- (39 years old) replies "*this is a big problem; we (Argentinean tourism scholars) are not updated to what is being published in the global north. The lack of proficiency in English is an obstacle, as well as the lack of access to leading indexed journals. English is the most spoken language in the industry, but paradoxically we –tourism researchers- have serious problems to write a decent paper in English, a paper that meets standard qualities ... see Scopus and tell me the gravitation of Argentineans to knowledge production. This causes distress for tourism education since we give the same bibliography as our professors. In consequence, there is a gap between theory –fully not updated- and practice (the problems of the tourism industry today). I mean the only person who gained prestigious in the tourism fields was Regina Schluter, Maximiliano Korstanje, Roberto Bertoncello and Elisa Pastoriza are great scholars too... But we need more researchers in the same direction. The lack of access to international tourism research led us to myopia that affects how syllabuses are designed. I feel Douglas Pearce has the right the research should be anglicised to be more professional. A couple of years ago I read a paper about this".*

In the above-referred excerpt, Roberto laments the exclusion of Argentinean tourism researchers of global knowledge production as well as their absence at leading publications. Basically, this point leads to an inevitable stagnation and closure to new theories. Editorial indexes are vehicles towards the professionalization of tourism research in Argentina, so to speak the only way to update syllabuses and tourism education.

4.6 Future Research Lines

To think the future research lines of tourism and hospitality in a post-COVID-19 context is very hard to grasp, simply because the future of tourism seems to be a bit uncertain. In fact, in this post-COVID-19 world, new technologies lead us to a landscape without tourists. It is difficult to resist the impression that digital technology helps in the creation of more resilient and smarter destinations (Sigala, 2020). Of course, the impacts of COVID19 are different in the Global North than in the South. The pandemic has ushered under-developing economies into an unparalleled economic downturn whilst only a few nations will access the vaccine. Those economies dependant on tourism will face devastating effects in comparison to those which combine tourism with other subsectors. Having said this, the present study case – with a focus on Argentina- offers two important assumptions. On one hand, the reviewed syllabuses are formed by tourism marketing and management courses but they are not updated to the new times. Given this, participants criticize the lack of training of the tourism professionals as well as post-graduate students. Besides, tourism students are not being educated as scientists well-versed to conduct applied research, as interviewees adhere. On another, participants refer to the exclusion of Argentina from the global knowledge production. Participants believe that the lack of interest in formal education for resilient destination explains by the exclusion of Argentinean academicians from the leading top-ranked journals. This creates a dissociation between theory and practice.

4.7 Conclusion

The COVID-19 affects negatively the tourism industry whilst revolting tourism education as never before. The urgency of training new policy-makers specialised in post-disaster recovery facets has been captivated the attention of scholars worldwide. New times call for new epistemologies, methodologies and transdisciplinary research to resolve unparalleled problems (Wen et al., 2020). The present study case, which is focused on Argentina, reflects the gap between theory and Practice in tourism education. Per our review, courses and tourism syllabuses lack of courses oriented to post-disaster marketing or governability. In the same way, some interviewees manifest their worries the current tourism research and syllabuses are not updated in Argentina. Whereas some experts agree the lack of access to leading journals or the linguistic barriers would be a credible reason, others toy with the belief that the syllabuses content was cloned from the private sector in the 80s. As a result of this, courses are based on an old-fashioned economic-centered paradigm that excludes post-disaster marketing, governability or even risk-management. The limitation of the precautionary doctrine, in which future risks can be forecasted and eradicated, is mentioned by one of the participants. Lastly, study-case shows two important assumptions. On one hand, the syllabuses should be adapted to new paradigms and contexts. On another, the idea of risk management is given space for resiliency (adaptation).

References

Airey, D., Tribe, J., Benckendorff, P., & Xiao, H. (2015). The managerial gaze: The long tail of tourism education and research. *Journal of Travel Research, 54*(2), 139–151.

Amaratunga, D., & Haigh, R. (2011). *Post-disaster reconstruction of the built environment: Rebuilding for resilience*. Wiley.

Amore, A., Prayag, G., & Hall, C. M. (2018). Conceptualizing destination resilience from a multi-level perspective. *Tourism Review International, 22*(3–4), 235–250.

Bass, R. V., & Good, J. W. (2004). Educare and educere: Is a balance possible in the educational system?. *The educational forum* (pp. 161–168). Taylor & Francis Group.

Benckendorff, P., Moscardo, G., & Pendergast, D. (Eds.). (2010). *Tourism and generation Y*. Wallingford.

Carr, A. (2020). COVID-19, indigenous peoples and tourism: A view from New Zealand. *Tourism Geographies, 22*(3), 491–502.

Chew, E. Y. T., & Jahari, S. A. (2014). Destination image as a mediator between perceived risks and revisit intention: A case of post-disaster Japan. *Tourism Management, 40*, 382–393.

Christou, E., & Sigala, M. (2001). Professional development in hospitality and tourism education: A strategy for the 21st century. *The International Journal of Tourism Research, 3*(4), 328–330.

Cochrane, J. (2010). The sphere of tourism resilience. *Tourism Recreation Research, 35*(2), 173–185.

Costa, C. (2001). An emerging tourism planning paradigm? A comparative analysis between town and tourism planning. *International Journal of Tourism Research, 6*(1), 425–441.

Gössling, S., Scott, D., & Hall, C. M. (2020). Pandemics, tourism and global change: A rapid assessment of COVID-19. *Journal of Sustainable Tourism, 29*(1), 1–20.

Gotham, K. F. (2017). Touristic disaster: Spectacle and recovery in post-Katrina New Orleans. *Geoforum, 86*(1), 127–135.

Gretzel, U., & Scarpino-Johns, M. (2017). Destination resilience and smart tourism destinations. *Tourism Review International, 22*(3–4), 263–276.

Gretzel, U., Fuchs, M., Baggio, R., Hoepken, W., Law, R., Neidhardt, J., et al. (2020). E-tourism beyond COVID-19: A call for transformative research. *Information Technology & Tourism, 22*(1), 187–203.

Hall, C. M., Prayag, G., & Amore, A. (2017). *Tourism and resilience: Individual, organisational and destination perspectives*. Channel View Publications.

Hayes, S. (2020). Socialising students for philosophic practice? An analysis of learning outcomes in tourism taught Master's programmes. *Journal of Hospitality, Leisure, Sport & Tourism Education*. ahead of print DOI: 100274.

Haywood, K. M. (2020). A post COVID-19 future-tourism re-imagined and re-enabled. *Tourism Geographies, 22*(3), 599–609.

Higgins-Desbiolles, F. (2020a). Socialising tourism for social and ecological justice after COVID-19. *Tourism Geographies, 22*(3), 610–623.

Higgins-Desbiolles, F. (2020b). The "war over tourism": Challenges to sustainable tourism in the tourism academy after COVID-19. *Journal of Sustainable Tourism*. ahead of print, 1–19.

Inui, Y., Wheeler, D., & Lankford, S. (2006). Rethinking tourism education: What should schools teach. *Journal of Hospitality, Leisure, Sport and Tourism Education, 5*(2), 25–35.

Ivanov, S., & Webster, C. (2019). Conceptual framework of the use of robots, artificial intelligence and service automation in travel, tourism, and hospitality companies. In *Robots, artificial intelligence, and service automation in travel, tourism and hospitality*. Emerald Publishing Limited.

Ivanov, S., & Webster, C. (2020). Robots in tourism: A research agenda for tourism economics. *Tourism Economics, 26*(7), 1065–1085.
Jacobsen, J. K. S., & Munar, A. M. (2012). Tourist information search and destination choice in a digital age. *Tourism Management Perspectives, 1*(1), 39–47.
Jennings, G. (2001). *Tourism research*. John Wiley and sons Australia, Ltd..
Jeong, M., & Shin, H. H. (2020). Tourists' experiences with smart tourism technology at smart destinations and their behavior intentions. *Journal of Travel Research, 59*(8), 1464–1477.
Kapucu, N., & Khosa, S. (2013). Disaster resiliency and culture of preparedness for university and college campuses. *Administration & Society, 45*(1), 3–37.
Korstanje, M. E. (2016). *The rise of Thana capitalism and tourism*. Routledge.
Korstanje, M., & George, B. (2020). *The nature and future of tourism: A post COVID19 context*. Apple Academic Press.
Korstanje, M. E., & Ivanov, S. H. (2012). Tourism as a form of new psychological resilience: The inception of dark tourism. *CULTUr-revista de Cultura e Turismo, 6*(4), 56–71.
Korstanje, M. E., & Seraphin, H. (2018). Awakening: A critical discussion of the role of robots in the rite of hospitality. In *Tourism and hospitality: Perspectives, opportunities and challenges* (pp. 59–77). Nova Science Publishers.
Lew, A. A., Cheer, J. M., Haywood, M., Brouder, P., & Salazar, N. B. (2020). Visions of travel and tourism after the global COVID-19 transformation of 2020. *Ahead of print DOI*. https://doi.org/10.1080/14616688.2020.1770326
Light, D. (2017). Progress in dark tourism and thanatourism research: An uneasy relationship with heritage tourism. *Tourism Management, 61*(1), 275–301.
Lin, Y., Kelemen, M., & Tresidder, R. (2018). Post-disaster tourism: Building resilience through community-led approaches in the aftermath of the 2011 disasters in Japan. *Journal of Sustainable Tourism, 26*(10), 1766–1783.
Nepal, S. K. (2020). Adventure travel and tourism after COVID-19–business as usual or opportunity to reset? *Tourism Geographies, 22*(3), 646–650.
Omar Parvez, M., & Cobanoglu, C. (2021). Opportunities and challenges of utilizing service robots in tourism industry: A tool for recovery from COVID-19 pandemic. *Journal of Smart Tourism, 1*(3), 17–20.
Paris, C. M. (2015). Social constructivism and tourism education. *Journal of Hospitality, Leisure, Sports and Tourism Education (Pre-2012), 10*(2), 103–115.
Prayag, G. (2020). Time for reset? COVID-19 and tourism resilience. *Tourism Review International, 24*(2–3), 179–184.
Qiu, H., Li, Q., & Li, C. (2020). How technology facilitates tourism education in COVID-19: Case study of Nankai University. *Journal of Hospitality, Leisure, Sport & Tourism Education*. ahead of print DOI: 100288.
Ratten, V., & Jones, P. (2020). Covid-19 and entrepreneurship education: Implications for advancing research and practice. *The International Journal of Management Education*. ahead of print, 100432.
Robbie, D. (2008). Touring Katrina: Authentic identities and disaster tourism in New Orleans. *Journal of Heritage Tourism, 3*(4), 257–266.
Ruhanen, L. (2006). Bridging the divide between theory and practice: Experiential learning approaches for tourism and hospitality management education. *Journal of Teaching in Travel & Tourism, 5*(4), 33–51.
Sather-Wagstaff, J. (2016). *Heritage that hurts: Tourists in the memoryscapes of September 11*. Routledge.
Séraphin, H., Zaman, M., & Fotiadis, A. (2019). Challenging the negative image of postcolonial, post-conflict and post-disaster destinations using events: The case of Haiti. *Caribbean Quarterly, 65*(1), 88–112.
Seraphin, H., Korstanje, M., & Gowreesunkar, V. (2020). Diaspora and ambidextrous management of tourism in post-colonial, post-conflict and post-disaster destinations. *Journal of Tourism and Cultural Change, 18*(2), 13–132.

Sigala, M. (2020). Tourism and COVID-19: Impacts and implications for advancing and resetting industry and research. *Journal of Business Research, 117*(1), 312–321.

Sunstein, C. R. (2005). *Laws of fear: Beyond the precautionary principle*. Cambridge University Press.

Tiwari, P., Séraphin, H., & Chowdhary, N. R. (2020). Impacts of COVID-19 on tourism education: Analysis and perspectives. *Journal of Teaching in Travel & Tourism*. ahead of print.

Tribe, J. (1997). The indiscipline of tourism. *Annals of Tourism Research, 24*(3), 638–657.

Tribe, J. (2000). Indisciplined and unsubstantiated. *Annals of Tourism Research, 27*(3), 809–813.

Tribe, J. (2010). Tribes, territories and networks in the tourism academy. *Annals of Tourism Research, 37*(1), 7–33.

Tribe, J., & Liburd, J. J. (2016). The tourism knowledge system. *Annals of Tourism Research, 57*(1), 44–61.

Tucker, H., Shelton, E. J., & Bae, H. (2017). Post-disaster tourism: Towards a tourism of transition. *Tourist Studies, 17*(3), 306–327.

Tyrrell, T. J., & Johnston, R. J. (2008). Tourism sustainability, resiliency and dynamics: Towards a more comprehensive perspective. *Tourism and Hospitality Research, 8*(1), 14–24.

Tzanelli, R. (2016). *Thanatourism and cinematic representations of risk: Screening the end of tourism*. Routledge.

Walters, G., Mair, J., & Lim, J. (2016). *Sensationalist media reporting of disastrous events: Implications for tourism. Journal of Hospitality and Tourism Management, 28*, 3–10.

Wang, J., Ayres, H., & Huyton, J. (2010). Is tourism education meeting the needs of the tourism industry? An Australian case study. *Journal of Hospitality & Tourism Education, 22*(1), 8–14.

Webster, C., & Ivanov, S. (2020). *Robots in travel, tourism and hospitality: Key findings from a global study*. Varna.

Wen, J., Wang, W., Kozak, M., Liu, X., & Hou, H. (2020). Many brains are better than one: The importance of interdisciplinary studies on COVID-19 in and beyond tourism. *Tourism Recreation Research.*, ahead of print, 1–4.

Xiang, Z., Tussyadiah, I., & Buhalis, D. (2015). Smart destinations: Foundations, analytics, and applications. *Journal of Destination Marketing and Management, 4*(3), 143–144.

Xiao, H., & Smith, S. L. (2010). Professional communication in an applied tourism research community. *Tourism Management, 31*(3), 402–411.

Zeng, Z., Chen, P. J., & Lew, A. A. (2020). From high-touch to high-tech: COVID-19 drives robotics adoption. *Tourism Geographies, 22*(3), 724–734.

Chapter 5
PANCOE a Method Applied on Joy Labs, University of Palermo, Argentina

Abstract Over years, the higher tourism education has centred on Cartesian Dualism which entailed a dissociation between rationality and emotionality. At the same time, pleasure was seen as a negative aspect to tourism education. PANCOE is the name of a project, based on Joy Labs at the University of Palermo, oriented to potentiate students' skills and performance through pleasure and joy. Echoing Epicurean legacy, PANCOE presents as an innovative and trailblazing project to improve tourism education. The goals of the project were twofold. On one hand, PANCOE uses digital technologies to create a friendly climate to homogenise academic grades in students coming from neighbour countries, but what is more important, PANCOE showed to be an efficient instrument not only to engage with students but also in reducing drop-out rates.

Keywords PANCOE · Pleasure · Education · Degree · Grades · Drop-out rates

5.1 Introduction

Some critical studies have cautioned the paradox of tourism research. This paradox lies in the fact that beyond the discipline maturation considerable knowledge fragmentation prevails. This fragmentation, as it was debated in earlier parts of this book, has invariably led to a deep crisis in higher tourism education. It is not simplistic to say that the classic curricula, which engages with the managerial perspective, fails to give coherent training to the next workforce. The rise of global dangers which threatens the industry as never before, such as terrorism, the ecological crisis, political violence without mentioning natural disasters, force educators to adopt radical shifts. In this vein, education offers a fertile ground for policymakers to find a creative alternative to resolve the problems of the sector (Ayikoru et al., 2009; Farmaki, 2018; Ndou et al., 2019; Sheldon et al., 2011). The problem of sustainability has marked agenda not only for policy-makers but also for scholars worldwide (Wilson & der Heidt von, 2016). To all these problems adds the lack of updating in the curriculum, as well as the apathy of students for the syllabuses and

A. Zuccoli, M. E. Korstanje, *The Role of Pleasure to Improve Tourism Education*, https://doi.org/10.1007/978-3-031-21580-3_5

curricula content, has been debated in the former book chapters. In these sections, we described the post-graduate frustrations as linked to the excessive working hours, and the low wages paid to front-desk staff (Korstanje & George, 2021).

As the previous backdrop, students´ psychological frustrations crystallise not only in the higher dropout rates but also in the low degree and academic performances just after the exams (Benjamin & Kline, 2019; Hsu, 2013; Korstanje, 2008). Some voices have lamented the bad experience of students after international internships resulted from linguistic barriers, xenophobia or simply discontent of employers with students´ performance (Cai et al., 2015; De Guzman et al., 2020). Although these studies illustrate the macro-structural limitations of students´ motivation less attention was paid to the role played by emotionality in the process. This means that educators and policymakers preferably focus on macro-structural variables such as curriculum, private funding or the global forces that mould higher education instead of delving into the students´ inner world. To fill the gap, the present chapter brings some reflections and obtained outcome of a new (successful) experiment with basis on the Joy Labs (at the University of Palermo, Argentina). This experiment was originally drawn and aimed at potentiating the student's performance whilst reducing the dropout rates. In so doing, the experiment combined digital technologies, with gamification theory through the creation of pleasurable experiences. As declared in the introductory sections, we hold the thesis that pleasure –far from being a disruptive emotion- enhances the quality of tourism education. PANCOE is a project aimed at stimulating the learning skills of the first undergraduate (UG)) students reported on a Bachelor in Tourism and Hospitality Bachelors. Those participants taking part in this experiment came from neighbouring countries such as Ecuador, Bolivia, Brazil, Chile, Venezuela and Colombia (only to name a few). Described as a study case, this chapter stresses the role played by pleasure (joy) as an optimiser of learning abilities as well as skills whilst giving an innovative technique to overcome the current barriers in tourism education. Students were encouraged to engage with different team-building activities like baking pieces of bread, food tasting, and cooking. Centred on Cartesian dualism, the tourism curricula do not include these types of activities in programs. What is more important, the figure of pleasure is relegated to a marginal position in the higher education system. PANCOE situates as a pioneering program main based on the Joy Labs to motivate deprived and low-ranked students. The experiment was conducted on basis of two different samples (formed by 10–20 participants in each one). The main goal of PANCOE is strictly associated with the stimulation of creativity of students with the support of AI (Artificial intelligence) and the use of social digital networks such as Twitter. In addition, Joy Labs explores the importance of smart ideas created by the configuration of safe social networks. The study coins the term *collective intelligence* to denote the rapid optimization of learning derived from students´ mingling surpassing its capacities in exams and tests. The sample was drawn in two groups formed of 870 participants who regularly are familiar with Twitter (Twitter-@holapancoe). The sample splits into two, active and passive participants. Whereas the former signals to tourism undergraduate students who take part of PANCOE (almost 40 students), the latter refers to students coming from other universities or areas but excluded from the experiment (830 students). The involved age cohort oscillates from 18 to 26 years old. Originally, the sample was formed by students taking the course of Integracion emocional, ambientacion y communicacion (Communication, environmentalism and emotional integration). It is not otiose to say that PANCOE seeks to integrate the senses in a different direction directly with emotions and academic performance. Digital platforms occupy a central place transforming negative feelings like frustration and fear into positive ones (i.e. trust and joy). The main learning objectives are as follows:

- To accelerate the inclusion of foreign students who are joining a course as year 1 students in tourism and hospitality careers.
- To potentiate the student learning experience, alongside reducing the rate of students who drop out.
- To make effective innovative methods to improve traditional tourism education, using tourism marketing case studies

PANCOE is part of 18 encounters lasting 90 minutes each one. In turn, participants were pre-graduate students well-skilled in tourism marketing and management. PANCOE is not a method but a set of different techniques combined to stimulate students' academic performance whilst reducing the dropout rates. The study examines the figure of pleasure as an innovative strategy to fill the gaps in classic tourism higher education. The obtained results show not only who students who participated in PANCOE have higher degrees but also how endorphins liberated from positive and pleasurable feedback pave the way towards better academic performances. As a promising technique, PANCOE supports directly the learning process of foreign and deprived students whilst outcomes suggest its application in other new universes, for example in the post-COVID19 context.

In the constellations of tourism, many pre-graduate students simply fail to finish their careers or simply leave the university because of high competence and an emerging sentiment of psychological frustration. The literature suggests that once earned their degree the young professionals are subject to excessive working hours associated with low-paid salaries. There is a gap between students' ideals and the reality of the marketplace in the West. With basis on the importance of pleasure (wellness) to stimulate the student cognitive attention, PANCOE which is based on the Joy Labs (laboratory of Joy) at the University of Palermo, Argentina combines different techniques oriented to reduce the student drop-outs as well as improving academic performance. Having said this, PANCOE centres efforts on the role played by pleasure and wellness in the education process. Interesting lines of research can be applied –following this methodology- in Asian universities. PANCOE recycles a climate of extreme competence which is often a source of distress and frustration, in valuable cooperation (gamification) marked in a constructive cooperation process. It is safe to say that the figure of play is vital to understand the PANCOE's success.

5.2 Theoretical Discussion

As discussed in earlier chapters, the word education stems from Latin Educere which means not only a type of spiritual emancipation but also was historically utilised in ancient Rome to feed cattle. So to say, one might assume that education has an ideological imprint as an instrument towards domestication (domination). As discussed in several chapters of this book, education has been widely framed by the positivist paradigm which is mainly based on Cartesian Dualism. As a manifest gap between emotionality and the mind, this dualism has accompanied curriculum and higher tourism education from its outset (Rozemond, 1998). In consonance with this, Cartesian dualism, which dominated tourism curricula, looked to optimize students' rating and academic performance as the only valid form of motivation. In the fields of tourism, Cartesian dualism has taken the lead filling the curricula worldwide (Ayikoru et al., 2009; Sheldon et al., 2011; Su, 2014). It is not simplistic to say that educators worked hard to stimulate students' cognitive skills whilst incorporating innovative strategies to deal with the global risks that place the tourism industry in jeopardy. Unfortunately, traditional education evinced major problems to be reformulated to cope with these new global dangers (Lew 201); to wit, these new emerging global dangers include terrorism, climate change, the outbreak of viruses like COVID-19 and political violence only to name a few) (Bagri & Junaid, 2020; Korstanje & George, 2021; Sigala, 2020). In this respect, some studies emphasised the urgency to put tourism education as a significant object of study in the constellations of tourism research. At a closer look, education provides policymakers with a robust background to design their marketing campaigns and programs whilst shedding light on the methodological limitations of applied research and the economic-centred paradigm (Lewis & Tribe, 2002; Tribe, 2002a, 2002b). As a result of this, tourism education serves as a catalyst that shortens the gap between theory and practice. This begs a more than the vexed question: what is tourism education and in what way it can be defined?

Tourism education should be understood as a limonoid process self-oriented to train the next policymakers, practitioners and scientist researchers to solve the real problems of the industry. The learning process permits an efficient interaction amongst the involving stakeholders (Belhassen & Caton, 2011; Cooper, 2002; Paris, 2011). It is noteworthy that digital technologies are key forces in the process of training the future workforce. Based on a complex interaction, the customer's decision-making process is bombarded with a flux of mass information that should be ordered and ultimately digested. What is still salient, the state of hyper-competition –amongst destinations- is given by the multiplication of actors and dynamics. Scholars have called attention to the adoption of digital technologies as an instrument to forecast future scenarios and potential threats so that the integrity of the system not to be harmed (Buhalis & Law, 2008; Morellato, 2014). Beyond this problem, countless works have alerted on the fact digital technologies –in some conditions- affect the students´ attention in classrooms. On one hand, technology engenders standardised forms of knowledge which shared with the audience but-, on another, it undermines critical thinking ushering users into a stereotyped form of behaviour (Munar & Gyimóthy, 2013; Williams & Hobson, 1995). One of the main paradoxes of tourism education rests with the extensive publications situated in leading journals which does not suffice to give the industry of valid answers to articulate efficient solutions. A clear example of this seems to be the lack of familiarity of experts to predict the rise of COVID19 pandemic that finally placed tourism into the wall and the deep blue sea (Carr, 2020; Sigala, 2020). The crisis of tourism education should be understood in sync with other dynamics, some authors dubbed as "the end of tourism, at least as we know" (Carr, 2020; Dale & Hassanien, 2008; Korstanje & George, 2021; Hannam, 2009).

5.3 Tourism Education Crisis

As debated, tourism education is mainly marked by a dualism where emotions are certainly subordinated to reasoning. In this token, educators prioritise cognitive skills from the students´ emotionality. Not only tourism education is based on what A. Franklin named it as tourist-centricity", but also introduced the managerial perspective as a dominant paradigm (Franklin, 2014; Korstanje & George, 2021). Echoing this, some studies exert a caustic critique on the curricula formation, which is designed to educate future tourism practitioners instead of scientists. At the same, classic quantitative methods have displaced other techniques to a marginal position. For the sake of clarity, recently-emerged models associated with long-life learning have been systematically debarred from higher education (Cuffy et al., 2012; Tribe, 2002a, 2002b). From the beginning, tourism education –at universities and institutions- was aimed at giving an all-encompassing diagnosis to resolve the problems of the sector (Cooper & Shepherd, 1997; Jafari & Ritchie, 1981). As Airey and Tribe (2006) eloquently describe, the limitations of tourism epistemology can be solved whether educators escape to the classic theories and paradigms towards a process of professionalization. The lack of a coherent assessment leads policymakers into a conceptual gridlock. Furthermore, its efficiency is given when the goals of the industry are successfully met (Wang et al., 2010). Moving in levels of great uncertainness, the tourism industry operates through weak frameworks which are affected by the interaction and multiplication of countless actors as well as stakeholders protecting their particular gains and interests. Far from being a product for only one segment, tourism education should be tailored to countless actors. The main goal of tourism education is the combination of techniques and knowledge to adapt a product for different segments (Amoah & Baum, 1997). Tourism education is culturally enrooted in the fields of expectative and experiences. Having said this, Sheldon et al. (2011) hold that Tourism Education Future Initiative (TEFI) auguries a new epoch to assist educators with an all-pervading framework accelerating a radical change to give responses to the incoming global challenges the industry is going through. TEFI's ideals are centred on the need to define five mainstream values for education: stewardship, ethics, knowledge, mutuality and professionalism. Ethics is understood as the interplay between actions striving towards a communal good and personal behaviour. Complementarily, knowledge derives from the process of education where personal skills are developed. Whilst stewardship signals the responsibility to care for others maximising gains for the next generations, professionalism refers to the capacity to acquire specific skills and abilities to engage with guests´ satisfaction. Lastly, mutualism is defined as the respect involving different levels of the organization, which certainly includes self-awareness, open-mindedness, empowerment and critical reasoning. These five pillars inspire a solid background in tourism educators. Some studies have recently brought a critical ground to adopt new techniques and methods in tourism education. Academia should develop a more critical pedagogy to mould a better version of the curricula in three directions: individual freedom, social justice and business productivity. Although one must add, technology potentiates education to some extent, no less true seems to be that it recreates the conditions to a technocratic and standardised form of knowledge affecting seriously the curricula and students´ performance. Putting this in other terms, tourism education not only struggles for a more sustainable industry but also reverses the negative effects of industry in society (Belhassen & Caton, 2011; Gupta, 2016). Leadership provides future workforce to adopt innovative techniques and self-resolving methods that helps policy makers to deal with global risks (Dredge et al., 2015; Sheldon et al., 2009). Leadership is essential not only to introduce necessary changes in the industry but also to reach a transactional climate of cooperation and mutual assistance. Dredge and Schott (2016) point out that leadership can be understood as a broad process generated by a mobilising force that coordinates successfully human capital, infrastructure, skills and resources. Leadership needs from three elements: leaders and followers, communication, and a set of actions towards specific goals. One of the challenges of higher education aims to think in jobs which are unknown for all us today. The change of labour market associated to the increase of competence often presses policy-makers to imagine for innovative solutions to the problems of the industry. The tourism research has debated between two contrasting poles. The dominance of management and marketing speaks us of tourism as a growing industry resulted from the technological breakthrough. At the same time, social sciences allude to tourism as a social institution that keeps society working. The academic identity moves in the basic sense of integrating both visions into applied research.

Ultimately, the innovative techniques not only fix the current inequalities of the capitalist system but also sanitise the asymmetries given by the tourism development (Gupta, 2016; Inui et al., 2006). Hence, the vocational perspective deserves further discussions in the years to come. Many young students are pushed to be enrolled in tourism courses simply because they look for a rapid and safe path to get a job, but this rarely happens in reality once their degree is earned. This occurs often in under-developed economies where tourism dependant is higher than other developed economies (Ndou et al., 2019). Tourism students –once end their career- are subject to countless problems and bad working conditions which include excessive working hours, low paid wages and labour exploitation. It is important not to lose sight of the fact that students are psychologically frustrated at the time they enter in the labour market (Farmaki, 2018; Korstanje 2008; Korstanje & George, 2021). Daniel et al. (2017) warn that students receive an education which is useless to be applicated in daily life. What is still important, they meet with countless difficulties with no opportunity for efficient resolution.

5.4 The Power of Wellness and Joy

For some reason which is very hard to discussed here, wellness has been systematically ignored in tourism studies. The concept proposes a wide range of subtypes such a medicine, public health, evasion, pleasure only to name a few (Sheldon & Bushell, 2009). Recently, some voices have enthusiastically applauded the role played by wellness-centred tourism in the sublimation of psychological discontent. These works overtly say that tourism should be understood as an instrument for human health and wellness (Norman & Pokorny, 2017; Vada et al., 2020). The distressing conditions of work, associated with day-to-day burdens, speak us of the need of adopting vacation travels for health and wellness enhancement (Holm et al., 2017; Lehto & Lehto, 2019). Although wellness tourism remains a captivating theme to be studied in the constellations of tourism research, probably considering wellness tourism as the most ancient form of tourism in the Roman Empire (Smith & Kelly, 2006), less attention was given to wellness applied to tourism education. Thal et al. (2021) hold that wellness should be incorporated to improve pre-graduate students´ competencies in tourism curricula. Based on empirical research with a focus on the Delphi method conducted in wellness travel agents, authors say that wellness not only would optimize students´ performance but also laying the foundations towards a new foundational curriculum for undergraduate students. The specialised literature associates wellness tourism only to the tourists´ experience (Luo et al., 2018) but it may be very well applied in other contexts. For example, in Asia interesting forms of wellness-related experiences which include Yoga meditation, wellness nutrition, as well as many other physical and psychological activities can be added to tourism education. Lastly, Prebežac et al. (2016) alert the importance to adopt new imperative changes in tourism curricula, above all because of the dramatic changes the industry is going through. At least, skills and knowledge in tourism students must be re-defined. For the next decades, students would look for jobs that do not exist in our present. What tourism scholars learnt seems to be obsolete for the years to come. What is more important, the global ecological crisis presses educators for developing more values-based educational programs. Emotions and psychology play a leading role in the configuration of the next 21 s century tourism curricula.

5.5 Methodological Discussion

As stated in several chapters of this book, Joy Labs is a pedagogic initiative of the University of Palermo, Argentina. The Joy Labs innovates incorporating digital technologies to the higher education system. PANCOE complementarily is an experiment conducted in graduate students in the tourism and hospitality carriers. The experiment is mainly based in a fluid dialogue an interaction with students through different activities. The sample was shaped by 870 students (15 males and 25 females) whose cohort ranges between 18 and 26 years old. The sample is divided in two: active and passive participants. The active group is formed by tourism undergraduate students (almost 40 participants) whereas the passive group is represented by 830 students (400 females and 430 males). Those students who take part of group A were subject to countless activities related to pleasurable experiences which include baking pieces of bread, drawing and painting, cooking and tasting different culinary dishes. The technique is mainly based on the gamification theory which encourages students to use digital platforms to share their culinary products with their followers. Participants compete directly to gain further scores according to the number of followers. The group B was subject to classic training and education. Ultimately, both groups have been tested to classic exams in the fields of history, tourism management, tourism marketing, geographies, sociology, psychology and statistics (all them are courses appertaining to their elective year). Since the sample is not statistically representative the obtained outcomes cannot be extrapolated to other universes. Additionally, since the large of group are disproportional, grade averages and other statistical inferences are not allowed. Last but not least, students come from middle class houses in which case the outcomes remain uncertain for low-class students.

5.6 PANCOE and the Joy Labs

5.6.1 A New Technique in Tourism Education

In this section, we describe the main obtained outcome of PANCOE, an experiment which is based on different digital platforms as Facebook and Twitter adjoined to the use of multiple learning techniques. PANCOE combines the use of the five human senses, smell, taste, touch, sight and hearing. Students who take part in PANCOE were systematically exposed to positive feedback whilst testing their cognitive skills. At a closer look, their skills have been a notably potentiated whilst stimulating pleasure (joy) in different encounters. Students eat some cultural dishes whilst interacting with peers. Joy Labs and PANCOE received much recognition which include the shortlist for the *Reimagine education awards QS Stars awards 2016 (Wharton College).* The Joy Labs is centred on fine cuisine made by well-known chefs specialising in gamification and neurosciences. Through the articulation of pleasurable experiences, the Joy Labs gives students all the necessary skills and knowledge to pass their exams with success. Alternating memorable and pleasurable experiences, PANCOE allows the realising of endorphins so that students can improve their cognitive performance in classrooms. Once they have taken their exams, participants were better scored than other students who have not participated in the experiment.

5.6.2 The Joy Labs in Action!

The experiment is based on two focus groups mentored by Alejandra Zuccoli. The focus groups are described below. The experience takes place during different days of the elective year whereas students make a fluid dialogue with other students during the meetings.

Participants are undergraduate tourism bachelor students at the University of Palermo, Buenos Aires, Argentina. Students, divided into two samples (20 each), are placed to work in isolation -at the first stage-. The first group A is formed by students who take part in PANCOE whilst the B group is subject to traditional programs. At a first glance, students in group A have been invited to bake pieces of bread as well other activities as cooking or sharing traditional dishes with peers. Each participant opened a Twitter Account and was asked to have 50–60 followers to share information about the dishes and cooking recipes. As Fig. 5.1 shows, these accounts were titled "PAN" followed by a nationality. Each Twitter account exhibited the dishes and cultural backgrounds of students as well as other preferences, hobbies or sports.

The interaction seeks to emphasise the importance of selling their products whilst gaining traction regarding other players. Centred on the benefits of gamification theory, students formed a ranking with the number of followers and offered dishes. This competition –far from being disruptive- enhanced solidarity and empathy with peers. In the second phase, participants are encouraged to express freely their emotions and feeling. They share not only their testimonies but also problems, hopes and fears in the previous stages to PANCOE. Additionally, students are tested through standardised and regularly-based exams for more than 2 hours. The results are shared 2 weeks later. PANCOE is a method that discerns the importance of pleasure (joy) as a factor towards knowledge stimulation and the learning processes optimization. Having said this, PANCOE derives from the Spanish term PAN (Pensamiento Automatico naturalizado -Automatic and naturalised thinking) and COE (Comunicacion basada en el estudiante- Communication based on students) (Fig. 5.2).

PANCOE looks to stimulate the endorphins in the brain exploring students´ skills as well as their creativity. Originally designed to potentiate the academic performance of deprived students or students coming from neighbouring countries who lived alone in Buenos Aires, PANCOE uses pleasure to overcome the barriers in

Fig. 5.1 PAN Alemania – Bread from Germany

classic education. Supported by digital technology, the experiment circumscribes to the durable state of pleasure marked by wellbeing and happiness.

Each participant offers a packaged product empowering the self-reliance to compete with others peers (to get more followers). The Twitter account is anonymous to bolster a frank and genuine dialogue with others. This helps participants to dismantle the proper emotional barriers. PANCOE is based on 18 encounters where students share their experiences with peers but also with well-renowned Argentinean chefs without mentioning students and lecturers from other faculties. Complementarily, in each encounter students, taste cultural dishes cooked by their classroom mates. This process is vital to enhance students´ interaction, allowing a rapid optimization of acquired learning and skills. Besides, the experiment re-integrates multiple sensory dimensions of students in a friendly-marked climate of cooperation. In each meeting, participants are asked to imagine a preferred tourist destination. At this stage, they are invited to express emotions on the experience in writing. Because of privacy and ethics, their names cannot be revealed. Maria,

Fig. 5.2 PAN El Salvador

(female, 22 years old) said *"I had a great experience with PANCOE. The method can be applied to everything I can do in life… it was as to apply everything we learned in the classroom but the real life. I did not how to create a Twitter account, I learned to use Twitter to upload my bread as well as sharing my experiences with unknown other students. I was shy to share my feelings, PANCOE was a great experience for me"*. Following this, Julieta (female, 20 years old) replies *"to my end, the PANCOE embodies directly into the communicative process; each student shares a product (a bread) to share a message with professors, students and other peoples. This denotes a circular model where we are receivers and senders. The experience questioned my inner-world whilst emotionally situating as a real challenge but it was indeed gratifying!"*

The joy labs kick off a significant change in classic tourism education, as Rosa notes in her interview: "*it was an uncanny experiment, creative, exhilarating, innovative and, different. Professor Alejandra offered me to use Twitter as an instrument of education and emancipation promoting our products to a vast public. Since the*

Twitter account was anonymous it was free to express my feelings. It was very interesting the previous moment to promote my slices of bread; I did not know who my followers were. My user-made public not only the loaves of bread but also fine gastronomy. Twitter serves as a fruitful and useful instrument to engage with others". In the same direction, we get with Rogelio (male, 20 years old) who overtly acknowledges that *"PANCOE is great! I never used Twitter; it takes some time to learn to go in twitter's world. Aside from this, I created my avatar; I presented my products whilst interacting positively with other users. This is the type of experience I wish to make again and again and again!*

In the reviewed specialised literature, the problems of students´ self-esteem and peer valorisation correlate directly with student dropout rates. PANCOE takes self-esteem as the main criterion to study as well Marcelo (male, 20 years old) responded *"I had a serious problem talking with others. This is my Achilles tendon, to the limitations I had because I am not familiar with Twitter, I am not in my native country, I am abroad out of my home. I am now learning to cook many dishes and I enjoyed getting new experiences during this experiment! The possibilities to cook my traditional dishes led me to my cultural roots.* Finally, Natalia (female, 18 years old) accepts *"PANCOE can be very well seen a new form of education simply because it put me in the Other's lens. In PANCOE I put myself in the other place to improve the relations as well as gain further understanding of the "Other". Using Twitter and PANCOE I unveiled my inner-world and my emotional barriers".*

Last but not least, the third stage is characterised by the application of regularly-based verbal and written exams; each exam lasted approximately 2 h. The outcomes matched the expectations as the students involved in the programme performed better than the others (the capacity of memory storage increased notably to 33% in students who participated in PANCOE comparing this with other groups who did not involve in the experiment (20%). Besides the number of students who graduated was 40% amongst those involved in the programme against 28% for the others (Fig. 5.3).

Amongst the methodology limitations of PANCOE, one must adhere that the investigation was strictly limited to middle-class foreign students –of course excluding Argentinean students. Hence, there is no evidence or results applied in lower-classes or groups aligned to durable psychological distress such as students coming from war zones, as well as countries marked by ongoing geopolitical conflicts or devastated areas. Let us share with readers that anyway promising approaches can be applied to higher tourism education just after PANCOE. It evinces an interesting potentiality to placate negative effects of COVID-19 in students situating as an innovative method that valorises the importance of pleasure, wellness and joy (Fig. 5.4).

Fig. 5.3 PAN Russia

5.6.3 *PANCOE in Perspective*

As mentioned in several parts of this chapter, PANCOE is an innovative method that incurs pleasure as a catalyst towards a better academic performance in pre-graduate tourism students. Methodologically speaking, the sample was drawn with two groups, A group which formed with students who took a personal experience with PANCOE, and a group B formed with students who did not take place in the experiment. Having said this, PANCOE has some strengths and weaknesses. Their results are not only promising but also categorical in the sense they exhibit a clear tendency. Students taking part in the experiment have not only better grades but also experienced fewer dropout rates. Amongst the weaknesses, PANCOE is very hard to be measured using the classic standardised pedagogical techniques. For that, we disposed of classic exams (equally applicable for those students in groups A and B). This means that students who were subject to PANCOE showed better cognitive

Fig. 5.4 Dishes and Magazines. (Source: Self-Elaboration, PANCOE Method, 2018)

skills in their exams than other samples. A second addition problem was the ratio marked by grade point average. In this respect, some commissions are former by larger than 50 students whilst others only are limited to 25 students. This suggests that ratios will considerably vary without showing a direct correlation between performance and learning skills. To solve this, we used a technique based on the average of students who successfully approved their elective year. Each elective year is based on 8 courses that should be successfully approved.

In 2015, 82% of students who have taken part in PANCOE (Group A) approved the elective year whilst only 68% of students (group B) passed the year. Three years later, in 2018 to be exact, Group A (PANCOE) successfully approved the year with 83% of their students in comparison with 69% in the group (B). The rise of the COVID-19 pandemic, doubtless, affected the performance of all students even in A group, but what is more important; for 2021, group A (Pancoe) showed how 75% of the students approved the year whereas only 67% applies for B group. All this evidence suggests important achievement in the fields of pedagogy with basis on PANCOE method. The results are detailed in Table 5.1.

The following figures shown below validate how effective PANCOE is even in times of COVID-19 (Figs. 5.5 and 5.6).

In group B the average rate never surpasses the 70% of students approving their respective elective year. The case is notably different with PANCOE who shows rates more than 80%.

Figure 5.7 represents the academic performance of B group only whereas comparing the higher and lower points of saturation of B group. This variance oscillates between 65 and 69% whilst PANCOE group ranges from 75 to 82%. Doubtless, this

Table 5.1 Performance of students in years 2015, 2018, 2021

Year	PANCOE - A	Group B
2015	82%	68%
2017	80%	65%
2018	83%	69%
2021	75%	67%

Source: self elaboration, PANCOE, 2012–2021

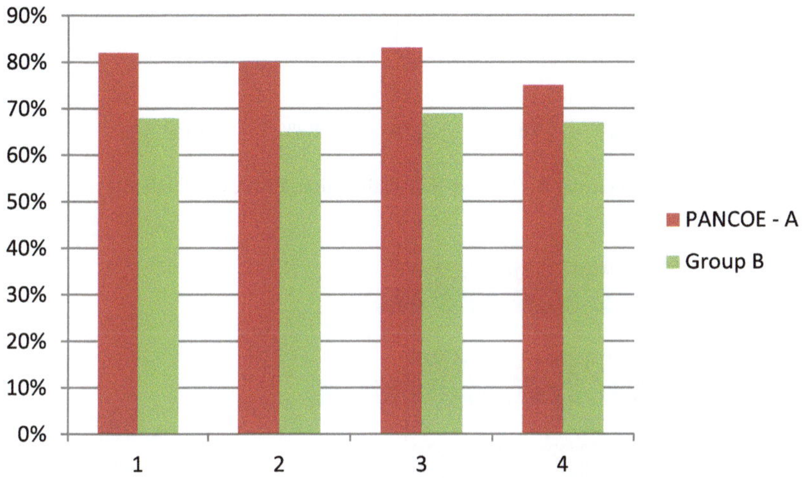

Fig. 5.5 Students Performance. (Source: self elaboration, PANCOE Method)

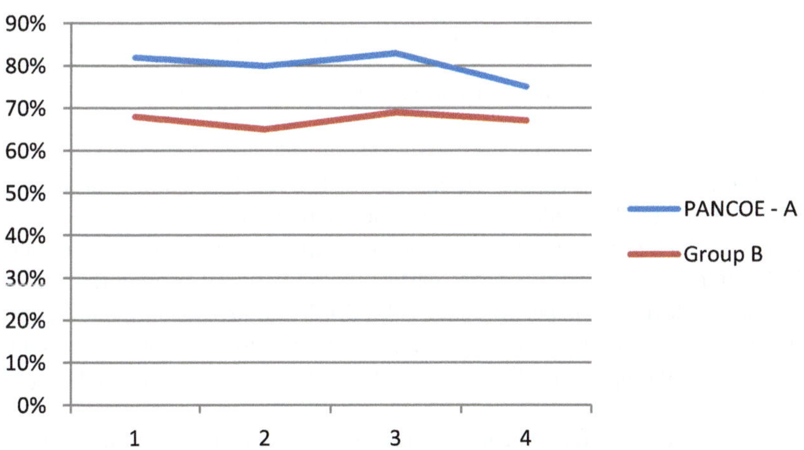

Fig. 5.6 Circular performance. (Source: Self elaboration, PANCOE method)

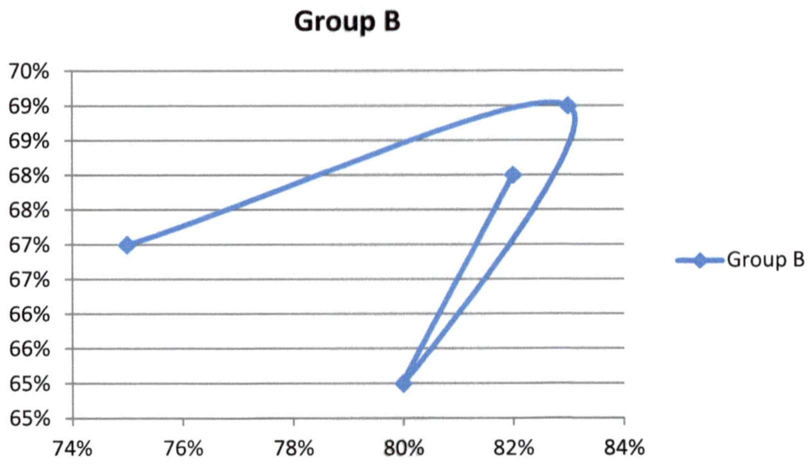

Fig. 5.7 Performance of Group B. (Source: Self-elaboration, PANCOE Method)

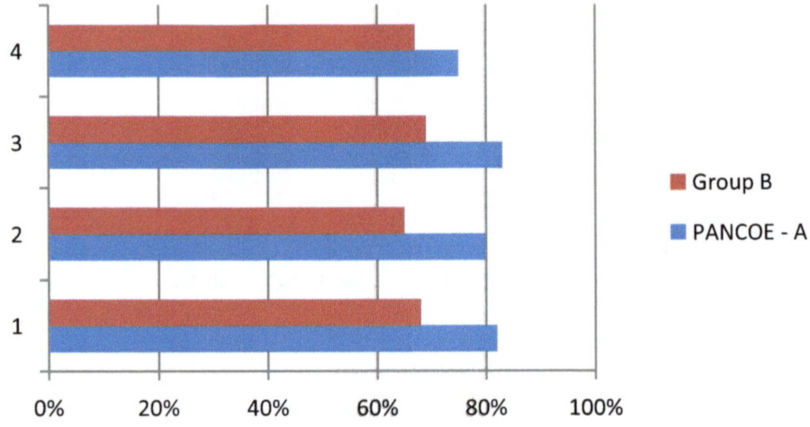

Fig. 5.8 Comparison Academic performance, Group A/B. (Source: self elaboration, PANCOE Method)

amply evidences not only the efficiency of PANCOE method but also how COVID-19 reduced the academic performance in both groups.

In Fig. 5.8 let's remind readers that each tier represents an elective year, tier 1 is for 2015, tier 2 is for 2018, 2 years earlier the COVID-19 pandemic. The tier 3 which notably decreases in performance is for 2021 year, a period fraught by restrictive measures and online-education. Despite the performance has been reduced, those students taking part of PANCOE have better grades than other groups (Fig. 5.9).

In the next section, we lay the latest comments and remarks revolving around the strengths and weaknesses of PANCOE method, as well as future applications for

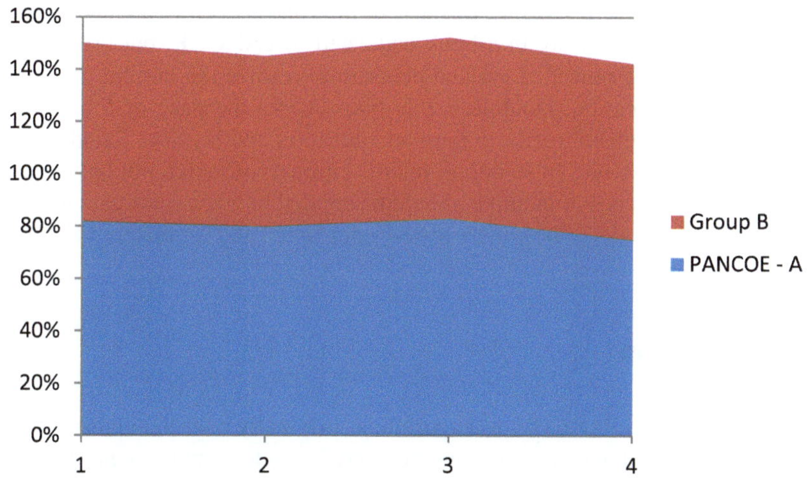

Fig. 5.9 Elective years approved. (Source: self elaboration, PANCOE method)

improving tourism higher education. In consonance with the findings obtained by Wellness tourism, we hold the thesis the figure of pleasure (joy) plays a leading role in expanding the current academic performance even in climate of extreme uncertainness and fear –like the COVID-19 pandemic period-.

5.7 Limitations of PANCOE and Future Research

As stated in several parts of this book, PANCOE is exploratory research that does not reach a statistically representative sample. Hence, even if promising the outcomes should be never extrapolated to other universes and contexts. Anyway, PANCOE has strengths in shedding the light on new [qualitative] approaches in the years to come. Having said this, qualitative approaches allow expanding the current understanding of the factors that impact [and determine] the current higher education crisis in the constellations of tourism and hospitality. Qualitative methods infer on the psychological frustrations students often come across, as well as other aspects which remain inexpungable to the researcher's eye. In consonance with this, PANCOE is not the only method which contributes to enhancing students' academic performance. We have discussed in the third chapter [to be more exact in the section entitled From the flow experience to joy] the FLIGBY method [as an experiential game] which helps potentiate students' creativity and adaptation in difficult times. To be best of our knowledge, gamification theory seems to be a fertile ground to optimize e-learning and virtual classrooms, in a post-COVID-19 context, as well as enhance academic engagement in frustrated students. FLIGBY articulates not only adaptative behaviour but also boosts students' resiliency to adapt to hostile situations. For that reason, some interesting inferences can be done between

FLIGBY and of course PANCOE. Because of time and space, we are unable now to develop this point, precisely in a book dedicated to PANCOE, but both methods centre mainly on the power of joy and pleasure to resolve the current problems in higher tourism education. Doubtless, this book marks the pace of a new line of inquiry that contributes directly to tourism education and lays the foundations for new models in the future. In so doing, interdisciplinary research not only is necessary but also these types of methods should be applied in other cultural contexts and country for their results to be widely tested, confronted and ultimately validated.

5.8 Conclusion

After further review, PANCOE -as experiment- originally instrumentalises the cognitive reserve of endorphins to engage poor-grade students but -sooner than later- the method was applied to other groups. To achieve safer social networks, the purpose of PANCOE links with the integration of creativity and Artificial intelligence and digital technologies. At the same time, PANCOE works actively to change negative emotions or experiences into joy (well-being). During the experiment, students were encouraged to dream with the best-preferred tourist destination sharing with other participants the culinary customs of their own country. Likewise, students were subject to watch pictures containing islands, beaches and paradisiacal landscapes. We denote the term imagined landscape to create a climate of well-being. The present book chapter reflected on the experience of PANCOE, an experiment conducted by the Joy Labs (University of Palermo, Argentina). The aims of this study were twofold. On one hand, it devotes efforts to introduce pleasure to reduce the current academic dropout rates. On another, it looks to standardise academic performance in low-graded or deprived students. Participants –in PANCOE- came from a neighbouring country –and lived far away from their friends and families-. Although classic tourism education has taken from Cartesian dualism its traction, this case goes in the opposite direction. PANCOE not only jolts classic education from any lingering complacency but also brings some reflection on the power of pleasure to potentiate students´ cognitive attention. The outcomes notably indicate that participants have better degrees than those who have no opportunity to take part of PANCOE. The endorphins liberated by positive feedback paved the way for new advances in gamification theory. Of course, more research is needed to understand how pleasure helps overcome the pitfalls of classic tourism higher education. Although promising the outcomes cannot be extrapolated to other universes or backgrounds simply because the sample was non-statistically representative. PANCOE, last but not least, evinces promising results even in contexts of extreme deprivation and distress like the COVID-19 pandemic. Of course, though the academic performance in students notably decreased in 2021 –in virtual online education- no less true seems to be that PANCOE group has better scoring than B group. This point, which is a deep-seated issue, should be re-approached in next layouts.

References

Airey, D., & Tribe, J. (Eds.). (2006). *An international handbook of tourism education*. Abingdon.

Amoah, V. A., & Baum, T. (1997). Tourism education: Policy versus practice. *International Journal of Contemporary Hospitality Management, 9*(1), 5–12.

Ayikoru, M., Tribe, J., & Airey, D. (2009). Reading tourism education: Neoliberalism unveiled. *Annals of Tourism Research, 36*(2), 191–221.

Bagri, S. C., & Junaid, K. C. (2020). Ethnic conflict and geopolitics: COVID-19 augmented chaos versus efforts to restart tourism sector. *Skyline Business Journal, 16*(1), 73–81.

Belhassen, Y., & Caton, K. (2011). On the need for critical pedagogy in tourism education. *Tourism Management, 32*(6), 1389–1396.

Benjamin, S., & Kline, C. (2019). How to yes-and: Using improvisational games to improv (e) communication, listening, and collaboration techniques in tourism and hospitality education. *Journal of Hospitality, Leisure, Sport & Tourism Education, 24*, 130–142.

Buhalis, D., & Law, R. (2008). Progress in information technology and tourism management: 20 years on and 10 years after the internet—The state of eTourism research. *Tourism Management, 29*(4), 609–623.

Cai, L. A., Wei, W., Lu, Y., & Day, J. J. (2015). College students' decision-making for study abroad–Anecdotes from a US hospitality and tourism internship program in China. *Journal of Teaching in Travel & Tourism, 15*(1), 48–73.

Carr, A. (2020). COVID-19, indigenous peoples and tourism: A view from New Zealand. *Tourism Geographies, 22*(3), 491–502.

Cooper, C. (2002). Curriculum planning for tourism education: From theory to practice. *Journal of Teaching in Travel & Tourism, 2*(1), 19–39.

Cooper, C., & Shepherd, R. (1997). The relationship between tourism education and the tourism industry: Implications for tourism education. *Tourism Recreation Research, 22*(1), 34–47.

Cuffy, V., Tribe, J., & Airey, D. (2012). Lifelong learning for tourism. *Annals of Tourism Research, 39*(3), 1402–1424.

Dale, C., & Hassanien, A. (2008). Podcasting in tourism education. *Journal of Teaching in Travel & Tourism, 7*(3), 43–60.

Daniel, A. D., Costa, R. A., Pita, M., & Costa, C. (2017). Tourism education: What about entrepreneurial skills? *Journal of Hospitality and Tourism Management, 30*, 65–72.

De Guzman, A., Bayot, J. M., Javier, N., Liamzon, R. R., & Moralejo, M. A. R. (2020). Understanding Filipino hospitality interns' expectations and frustrations. *Anatolia, 31*(3), 479–493.

Dredge, D., & Schott, C. (2016). Academic agency and leadership in tourism higher education. In D. Prebezac, C. Schott, & P. Sheldon (Eds.), *The tourism education future initiative: Activating change in tourism education* (pp. 36–60). Routledge.

Dredge, D., Schott, C., Daniele, R., Caton, K., Edelheim, J., & Munar, A. M. (2015). The tourism education futures initiative. *Anatolia, 26*(2), 340–346.

Farmaki, A. (2018). Tourism and hospitality internships: A prologue to career intentions? *Journal of Hospitality, Leisure, Sport & Tourism Education, 23*, 50–58.

Franklin, A. (2014). Tourist studies. In *The Routledge handbook of mobilities* (pp. 94–104). Routledge.

Gupta, V. (2016). Indian reality tourism-a critical perspective. *Tourism and hospitality management, 22*(2), 111–133.

Hannam, K. (2009). The end of tourism? Nomadology and the mobilities paradigm. In J. Tribe (Ed.), *Philosophical issues in tourism, 37* (pp. 101–115). Channel View.

Holm, M. R., Lugosi, P., Croes, R. R., & Torres, E. N. (2017). Risk-tourism, risk-taking and subjective well-being: A review and synthesis. *Tourism Management, 63*, 115–122.

Hsu, L. (2013). Work motivation, job burnout, and employment aspiration in hospitality and tourism students—An exploration using the self-determination theory. *Journal of Hospitality, Leisure, Sport & Tourism Education, 13*, 180–189.
Inui, Y., Wheeler, D., & Lankford, S. (2006). Rethinking tourism education: What should schools teach. *Journal of Hospitality, Leisure, Sport and Tourism Education, 5*(2), 25–35.
Jafari, J., & Ritchie, J. B. (1981). Toward a framework for tourism education: Problems and prospects. *Annals of Tourism Research, 8*(1), 13–34.
Korstanje, M. (2008). Education issues in tourism, an analysis of students expectatives in Argentina. *Tourism Today, 8*, 154–166.
Korstanje, M., & George, B. (2021). *The nature and future of tourism: A post COVID-19 context*. Apple Academic Press.
Lehto, X. Y., & Lehto, M. R. (2019). Vacation as a public health resource: Toward a wellness-centered tourism design approach. *Journal of Hospitality & Tourism Research, 43*(7), 935–960.
Lewis, A., & Tribe, J. (2002). Critical issues in the globalisation of tourism education. *Tourism Recreation Research, 27*(1), 13–20.
Luo, Y., Lanlung, C., Kim, E., Tang, L. R., & Song, S. M. (2018). Towards quality of life: The effects of the wellness tourism experience. *Journal of Travel & Tourism Marketing, 35*(4), 410–424.
Morellato, M. (2014). Digital competence in tourism education: Cooperative-experiential learning. *Journal of Teaching in Travel & Tourism, 14*(2), 184–209.
Munar, A. M., & Gyimóthy, S. (2013). Critical digital tourism studies. In *Tourism social media: Transformations in identity, community and culture* (pp. 245–262). Emerald Group Publishing Limited.
Ndou, V., Mele, G., & Del Vecchio, P. (2019). Entrepreneurship education in tourism: An investigation among European universities. *Journal of Hospitality, Leisure, Sport & Tourism Education, 25*, 100175.
Norman, A., & Pokorny, J. J. (2017). Meditation retreats: Spiritual tourism well-being interventions. *Tourism Management Perspectives, 24*, 201–207.
Paris, C. M. (2011). Social constructivism and tourism education. *Journal of Hospitality, Leisure, Sports and Tourism Education (Pre-2012), 10*(2), –103.
Prebežac, D., Schott, C., & Sheldon, P. (Eds.). (2016). *The tourism education futures initiative: Activating change in tourism education*. Abingdon.
Rozemond, M. (1998). *Descartes's dualism*. Harvard University Press.
Sheldon, P. J., & Bushell, R. (2009). Introduction to wellness and tourism. *Wellness and tourism: Mind, body, spirit, place*, 3–18.
Sheldon, P., Fesenmaier, D., & Tribe, J. (2009). The tourism education futures initiative. *E-Review of Tourism Research, 7*(3), 39–44.
Sheldon, P. J., Fesenmaier, D. R., & Tribe, J. (2011). The tourism education futures initiative (TEFI): Activating change in tourism education. *Journal of Teaching in Travel & Tourism, 11*(1), 2–23.
Sigala, M. (2020). Tourism and COVID-19: Impacts and implications for advancing and resetting industry and research. *Journal of Business Research, 117*, 312–321.
Smith, M., & Kelly, C. (2006). Wellness tourism. *Tourism Recreation Research, 31*(1), 1–4.
Su, Y. (2014). Lifelong learning in tourism education. *The Routledge Handbook of Tourism and Hospitality Education*, 322.
Thal, K. I., Smith, S. L., & George, B. (2021). Wellness tourism competences for curriculum development: A Delphi study. *Journal of Teaching in Travel & Tourism, 21*, 1–16.
Tribe, J. (2002a). Research trends and imperatives in tourism education. *Acta Turistica*, 61–81.
Tribe, J. (2002b). The philosophic practitioner. *Annals of Tourism Research, 29*(2), 338–357.
Vada, S., Prentice, C., Scott, N., & Hsiao, A. (2020). Positive psychology and tourist Well-being: A systematic literature review. *Tourism Management Perspectives, 33*, 100631.

Wang, J., Ayres, H., & Huyton, J. (2010). Is tourism education meeting the needs of the tourism industry? An Australian case study. *Journal of Hospitality & Tourism Education, 22*(1), 8–14.
Williams, P., & Hobson, J. P. (1995). Virtual reality and tourism: Fact or fantasy? *Tourism Management, 16*(6), 423–427.
Wilson, E., & der Heidt von, T. (2016). Business as usual? Barriers to education for sustainability in the tourism curriculum. In D. Prebezac, C. Schott, & P. Sheldon (Eds.), *The tourism education future initiative: Activating change in tourism education* (pp. 97–113). Routledge.

Chapter 6
The Future of Tourism Education Just after the COVID-19

Abstract It is almost impossible to discuss the future of tourism education without debating the post COVID-19 context. The industry of tourism not only was a perfect carrier for this virus but also its main victim. In some hindsight, COVID-19 led the industry on the bring of collapse but opening some doors to new opportunities in the fields of tourism education. The academic curricula are far from being updated. This chapter explores the limitations of current curricula and the opportunities for the next years. Most certainly, COVID-19 has triggered a hot debate in the Academia. Whilst some voices alert on the epistemological problem in studying tourism in a world without tourists, others see in the pandemic an opportunity to degrowth tourism. Quite aside from the discrepancy, this chapter sheds light on the potentialities of PANCOE to cope with frustrated students. Of course, since the sample is not statistically representative the findings should not be extrapolated to other universes.

Keywords Frustration · Fear · Pandemic · COVID-19 · Opportunities · Anxiety

6.1 Introduction

To close a book regarding tourism education seems to be very difficult without discussing the effects of the COVID-19 pandemic. The recent virus outbreak is known as SARSCOv2 (COVID-19), which was originally reported in Wuhan, China has infected more than 460 million and killed almost six million people giving a death blow to service sectors. To stop the pandemic, governments disposed of restrictive steps and protocols which included the closure of borders, airspaces and airports (Buckley, 2021; Korstanje, 2021). The tourism industry was not only the main carrier of the virus but also its main victim. The schools, universities and public buildings were closed as well as the adopting of restrictive measures to prevent daily contacts such as social distancing and a strict lockdown. All these measures not only affected global trade but also international and domestic tourism (Mao et al., 2021; Pappas, 2021; Sharma & Nicolau, 2020). Without any doubt, the COVID-19 pandemic not only altered the daily life in high-mobile societies but also the current

A. Zuccoli, M. E. Korstanje, *The Role of Pleasure to Improve Tourism Education*, https://doi.org/10.1007/978-3-031-21580-3_6

travel behaviour (Korstanje & George, 2021a, b; Tzanelli, 2022). Sharma and Nicolau (2020) argue convincingly that COVID-19 ushered the industry into an unparalleled devastating crisis. Beyond the health protocols different stakeholders have devoted efforts in planning the tourism recovery facet. Assessing the impacts of COVID-19 in different subsectors such as rental-cars, cruise, hotels and airlines industries, authors acknowledge that governmental intervention assisted great corporations but failed to provide financial aid to small corporations. In this vein, governmental interventions do not suffice to bail out corporations in the brink of bankruptcy. Zhang et al. (2021) question on the fact that the classic methods of forecasting to overcome the states of crisis in the past are now obsolete and old-fashioned techniques which should be improved. By evaluating the economic impacts of the pandemic in the industry, Zhang et al. said overtly that the re-branding process of destination is a key factor to boost domestic tourism. Fotiadis, Polyzos & Huan estimate that the tourism recovery should endure almost more than 12 months. Based on similarly-compared crisis scenario, authors lament the drop out in international tourism arrivals will persist in higher levels ranging from 30 to 70% even in 2021. Said this, the opinion of Academia has been divided into two clear-cut poles. On one hand, some scholars applauded enthusiastically the opportunity to lay a new conceptual basis of tourism towards a more sustainable form of consumption. For these voices, the ecological crisis opens the doors to a new low-mobile scenario where the effects of climate changes can be successfully reversed (Gössling et al., 2020; Prayag, 2020; Rastegar et al., 2021). On another, a dearth of studies emphasised the urgency to introduce changes in the tourism epistemology to foster an interdisciplinary method of study. Quite aside from this, the COVID-19 pandemic has not only long-lasting effects in the sector, but firms should operate in a climate of higher uncertainty (Korstanje & George, 2022; Wen et al., 2021). An emerging third way of scholars has alerted on the problems generated by COVID-19 in higher tourism education (Sigala, 2020; Tiwari et al., 2021). The adoption of AI (artificial intelligence) and digital technologies in virtual classrooms potentiated education in some direction but created a serious disruption in countries or cultures with low access to these technological platforms (Joshi & Gupta, 2021). In the mid of this mayhem, the current chapter explores the advantages and disadvantages of COVID-19 and its real impact on tourism higher education. At the same time, we elaborate on the challenges and problems of the sectors aggravated by the COVID-19 pandemic. Based on the PANCOE method, which was carefully described and explained in earlier sections, we elaborate an all-pervading model to mitigate the devastating aftermath of COVID-19 in students´ psychology. The COVID-19 pandemic represented a serious challenge for students and professors most probably aggravating the crisis of tourism education (Seo & Kim, 2021; Ye & Law, 2021a, b). The pandemic affected the life of almost 1.6 billons of learners globally. Students come across with serious mental distress resulted from the excess of online classes (Kallou & Kikilia, 2021; Seo & Kim, 2021). To some extent, the problems in tourism education are not news but the pandemic caused harms in the attractiveness of tourism-related careers (Reichenberger & Raymond, 2021). Let us explain readers that the point was widely addressed in previous chapters. The specialised literature enumerates a series of variables and key factors which correlate directly to the current tourism education crisis (Korstanje, 2021). Some of them are detailed below:

- There is a gap between theory and practice that leads towards a dissociation between what is taught in classrooms and the necessities of the industry (Ruhanen, 2006; Zhang et al., 2021)
- The hegemony of the economic-based paradigm excludes systematically other knowledge, methods and voices from the learning process (Airey et al., 2015)
- Digital technology often not only opens a chasm between those agents who have accessibility and those who are excluded but also affirms inter-class asymmetries or inequalities (Forristal, 2016; Liburd et al., 2016).
- Several graduate tourism students experience a profound state of frustration because of the inconsistencies and inhospitality of the labour market (Norrild & Korstanje, 2021).
- The curricula are not only far from being updated but also it is not adapted to the global dangers the industry should face (Chathoth & Sharma, 2007).

As the previous backdrop, Jafari and Ritchie (1981) have eloquently analysed that one of the limitations of education is associated with the lack of interest in educators for the external risks that place the industry in jeopardy. To this problem, they add the lack of interaction amongst stakeholders which often struggle to impose and protect their interests. The industry shows a complex dialogue amongst guests and hosts without mentioning sub-service sectors and exchangeable goods. Nowadays, it is important not to lose sight of the fact that the recent COVID-19 pandemic has generated devastating consequences for the tourism and hospitality industries affecting the operations of all sector levels (Zhong et al., 2021). In the former chapter, we have already presented the promising results of PANCOE applied during the pandemic; complementarily this book chapter offers practical and theoretical reasoning on how to cope with the tourism education problems in a post-COVID-19 world. Divided in two sections, the chapter interrogates furtherly on the effects of COVID19 in the industry and higher education system, whilst the second signals to the challenges and methodological limitations of PANCOE applied in times of social distancing and lockdown.

6.2 The Effects of COVID-19 in the Tourism Industry and Education

Experts agree that the COVID-19 pandemic which was originally reported in China generated serious and devastating consequences for the tourism industry (Haryanto, 2020; Hussain & Fusté-Forné, 2021; Koh, 2020) which encompass unemployment, bankruptcy and economic downturn, the fall of GDP of developed and underdeveloped economies without mentioning the export revenue falls (Bakar & Rosbi, 2020; Islam et al., 2020; Maital & Barzani, 2020). It is important to mention that these consequences varied on the levels of development of the affected country. As Rogerson & Rogerson noted, the economic effects of COVID-19 in the industry which include unemployment, economic downturn and degrowth tourism, have different implications in developed and underdeveloped countries. At the same time, those countries, which are dependants of tourism industry, experience an accelerated economic crisis in comparison with other mix-balanced economies (Rogerson et al., 2021; Rogerson & Rogerson, 2021a, b, c). To date, scholars have not reached any firm consensus that envisages the future of tourism in the years to come. This happens simply because the real effects of COVID-19 are uncertain even for experts (Romagosa, 2020). Collins-Kreiner and Ram (2020) pay the attention to the importance to draw national strategies to deal with the pandemic. The failure of government to placate the negative effects of COVID19 was certainly given by the low liquidity to bail out small companies as well as the dispersion of strategies in the national domain. In this sense, states provided little economic stimulus to accelerate the recovery of the sector. This begs a more than pungent question: is the COVID-19 the final executioner of tourism and global mobilities?

In a seminal book which entitles *Cultural immobilities and the virocene: mutating the crisis,* Professor Rodanthi Tzanelli (2021) explores the dilemmas of degrowth tourism which have been accelerated by COVID-19 pandemic. This virus marked a new epoch dotted with low-mobilities and great political fragmentation. Per her viewpoint, the tourism industry faces devastating consequences which mould a new emerging narrative based on the biomedical narrative. Of course, as she puts it, this narrative not only determines new types of travel behaviour but also a climate of uncertainness and unpredictability, where everything may happen anytime and anywhere. The restrictive measures –such as the social distancing- are contemplated as an ambiguous bridge which says who lives or are infected to die. In conclusion, COVID-19 brings some radical shifts which include a sense of immobility that chrysalises towards a sentiment of recrimination. This sentiment of recrimination, adjoined to conspiracy, leads us to a long dormant racism against Asian countries. This lack of mobilities is compensated by the obsession with digital technologies, as Tzanelli concludes. In the same reasoning, Mostafanezhad et al. (2020) document how in digital platforms users manifest an overt racism against Chinese tourists who are considered the carriers of the COVID-19 virus. These geopolitical anxieties coincide in what Korstanje and George (2021a, 2022) dubbed as "the re-feudalization of the world" which is certainly marked by the decline of hospitality in the West. Quite aside from this, some Chinese students have experienced acts of segregation and racism in Western countries during their internships (Li & Zhang, 2022; Wassler & Talarico, 2021; Wen et al., 2020). Korstanje and George (2021b, c, 2022) have called the attention to the state of crisis left by the COVID-19 pandemic. The turn of the twentieth century has brought not only many uncontemplated risks but also witnessed a climate of fragmentation as never before. In this token, the COVID-19 has ignited a process of feudalization mainly marked by geopolitical tensions, racism and chauvinist expression and of course a manifest hostility for foreign tourists. The question whether John Urry coined the term tourist-gaze has set the pace to a new "wicked-gaze" where tourists are catalogued as undesired guests. At least, the future of tourism industry is seriously compromised. Authors emphasise on the crisis of hospitality as a new sign of our days. The opposite is equally true, Chinese tourism students reconsidered their prone to study abroad (in the Western universities) because of the so-called hostility against foreigners in the mass media. In interesting study conducted by Cheng & Agyeiwaah demonstrate Chinese students` worries revolve around some points. The health risks to be infected and hospitalised in abroad -far away for China- occupied a central position of students` answers followed immediately by other fears such as discriminations in West universities, problems to engage with e-learning, and serious difficulties to find a top university to attend. What is more important, discrimination or segregation impedes students to make a genuine cross-cultural exchange (Cheng & Agyeiwaah, 2021). Wassler and Fan (2021) hold that pandemic has generated an unprecedented shock for the industry but what still seems to be unclear is the answer of the academia in this grim landscape. At a first glimpse, authors conduct qualitative-led interviews with 9 experts in the field who are geographically located in Europe, Asia and the Pacific region. Four items are evaluated by experts: tourism education, industry collaboration, applied research and discipline identity. Experts said that the figure of recovery is of paramount importance if the effects of COVID-19 are short-run but adaptancy should be incorporated if the effects are long-lasting. With this background in mind, Wassler & Fan emphasise the role of tourism academia which was empowered just after the crisis generated by COVID-19. As they go on to say *"Tourism academia is now facing an ambiguous situation, in which the phenomenon of reference (i.e. tourism) is suffering in an unprecedented manner in modern history. Some scholars have highlighted that it will be paramount for tourism academia to assist with research efforts during the crisis while discussions on academic networks (e.g. TRINET) have often raised a more general "quo Vadis" for the field"* (Wassler & Fan, 2021: 1). Following this reasoning, the academia debates between two contrasting tendencies: advocates of sustainability who see in the crisis an opportunity to re-birth, and advocates of industry who need urgent answers to accelerate the tourism recovery facet. Having said this, one of the main aspects that distinguish this from other pandemics namely SARS or Swine flu appears to be that the industry ground into an unparalleled halt, of course affecting countries in many but different ways. COVID-19 and its social impacts varied on the country which adopted different strategies. Just after operating with the involving variables, Wassler & Fan find that interviewed academicians manifest their worries on the major economic impacts on the COVID-19 over the industry as well as other derived subsectors. Other worries associated with political factors such as geopolitical tensions, or border closures are mentioned too. What is more important, technical factors that determine new e-learning and major issues in education which remould the discipline identity have been placed into the foreground for the consulted scholars. The priority of the Academia seems to be given in the urgency to fill the gap between theory and practice, so to speak, in the same ways, other disciplines filled the same gap. In so doing, the curricula should be substantially changed and adapted to these hard times. Edelheim (2020) discusses critically tourism education keeps an outstanding opportunity to introduce greater transformations in the sector, but to some extent, higher education overlooks the role played by research prioritizing business and profits. Historically universities were framed and designed to educate professionals, not researchers, as Edelheim eloquently observes. Tourism education overvalorises the economic values over other inspirational ones.

> Transformation of THE higher education is only possible if we start to question how and what it is that we value in concepts such as 'growth', 'greed', 'development', 'globalization', or 'climate change'. None of these concepts are value-free, quite the opposite, they are value-laden, and they shape the way we go about our daily lives, or how we create our professional identities. COVID-19 might be a blessing in disguise again here, as transformation of THE as a sector in the longer term is only possible if we equip our students, through the education we provide them, with tools not only to transform themselves, but also to transform the realities they will inhabit. The wonderful concept of worldmaking (Edelheim, 2020: 553).

As long as the restrictive measures take place, some scholars applauded the role played by digital technologies to reduce the negative effects on the higher education system (Sigala, 2020). In this vein, the evidence suggests that in developed countries with higher access to technologies students engaged with their classrooms whilst in under-developed countries the dropout rates notably halved (Amin et al., 2021; Qiu et al., 2021). Ratten and Jones (2021) discuss to what extent COVID-19 accelerated a crisis or potentiated new innovative learning skills in students. Entrepreneurship is vital to understanding these opportunities in a post-COVID-19 world. Per authors, since many things have changed after COVID-19, a new curriculum should not only reflect new entrepreneur content but also innovative guidelines mainly oriented to solve the real problems of the industry. In consonance with this, Amin et al. (2021) acknowledge that digital technologies correlate positively in the students´ self-efficacy and flexibility to new situations. The employment of digital technologies and informational technologies helped in the prevention of infections whilst boosting virtual interaction among students. In this way, e-learning, which was in use earlier in the pandemic, received further attention in western universities.

As the previous argument is given, Tiwari et al. (2021) enthusiastically toy with the belief that digital technologies help the educator to re-invent the syllabuses in what they dubbed as *ambidextrous management,* which means the delivery of careers overt to follow a variety of modalities in classrooms. This suggests a philosophical dilemma revolving around the fact that those establishments that fail to adopt innovation will invariably perish, as authors conclude. As Edelheim (2020) puts it, it is difficult to resist the impression that for a successful recovery of the sector, education should gain traction to change the cultural values of society. The COVID-19 not only woke up the industry from the slumber it was but also accelerated radical shifts towards a more sustainable form of consumption. In doing so, the old-fashioned cultural values of neoliberalism should be at least replaced. Citing Edelheim's reflections, Higgins-Desbiolles speaks to us of "a more ethical tourism", a point well reviewed in the former chapters. As she writes, in different of her texts, COVID-19 greases the rails of a new type of consumption and travel behaviour, with more empathy with the current ecological crisis. She uses Jim Butcher's term "the war over-tourism" to denote the binary between radical reform and tourism recovery. Per her stance, the crisis left by the pandemic has a major implication for a rebirth of the sector, a new rebirth more associated with sustainability and ecological justice. Over decades, the policies emanated from the WTO and other official organizations supported the neoliberal agenda, accelerating not only mobilities but also depredating the natural resources without remorse, in this context, tourism has directly or not supported the social injustices of the neoliberal agenda. Centred as a common good, COVID19 gives the opportunity to re-frame new community-based tourism for the next years (Higgins-Desbiolles, 2020, 2020b; Higgins-Desbiolles et al., 2021). Offering a transformational moment towards local tourism, policy-makers should work hard to include the local turn (ecology and environment) in their debates (Higgins-Desbiolles & Bigby, 2022). Contrariwise, other critical works are far from believing COVID-19 represents a founding point towards a new more ecological society. Instead, they speak us of a "fractured world" mainly marked by political violence, racism, segregation and geopolitical disputes without mentioning serious travel bans unilaterally imposed on strangers. In this untrammelled discourse, tourists are undesired guests -who equated to terrorists or carriers of a lethal virus-. This is a new normal where hospitality is simply dying. It is difficult to resist the impression something will change in this scenario simply because the current means of production and consumption are not changed –even just after the pandemic. Whilst some industries –like tourism- are in an irreversible decline, others like digital technologies have notably increased. Possibly, the capital flow has transformed from a global to a local perspective, but this does mean the birth of a new more ecological society. The question of whether J. Urry envisaged the tourist-gaze as the corollary triumph of globalization and tourist consumption, we coin the term "wicked gaze" to symbolise the new hostility of locals to foreign tourists (Banerjee, 2021; Barbosa et al., 2021; Korstanje, 2020; Korstanje & George, 2021c, 2022). Mohanty et al. (2020) argue that the social distancing, and online classrooms have been welcomed instruments which lasted considerable time after COVID-19. Artificial intelligence not only enriches the students` experience but also revitalises the conflicts between professors and students. AI contributes to the creation of more sustainable and safer post-COVID19 tourism environment. Educational institutions should reconsider the potentialities of AI in giving virtual trips to students in which case, whilst alleviating the pollution and contamination. Having said this, Ye and Law (2021a, b) hold more than an interesting thesis. They recognize the long-lasting psychological effects of COVID-19 in lecturers and students. To reverse the negative impacts, educators should formulate new policies, tests and exams integrating various types of knowledge and methods towards a multidisciplinary agenda. The classic evaluation and tests should not be applicable in a post-COVID-19 landscape. Echoing this, Nair (2021) stresses that the gamification theory offers a fertile ground to combine a sane competition with students´ engagement in the e-learning process. As she understands, no less true is that gamification has advantages and disadvantages. Nair mentions some benefits of gamification which includes the importance to respect "human diversity" and alterity, and enhancing further exams that motivate students. To wit, the figure of engagement occupies a central position in tourism education. Following the same reasoning, Zhong et al. (2021) draw attention to the role of engagement to placate students´ frustration and discontent whilst boosting their loyalties to the institution. Despite the pandemic, educators should think about the optimization of students´ motivation and direct engagement. COVID-19 has put the industry on the brink of collapse, affecting seriously the students´ expectations and dreams in the long run. With the benefits of perspective, digital technologies have not sufficed to potentiate education after the pandemic. In fact, in those cases where positive emotions were introduced such as joy, hope and wellness, digital technologies contributed positively to e-learning, but the opposite was equally true, in those cases where students experienced uncertainty, fear or anger, the obtained outcomes lacked being desired. This happens simply because positive emotions pave the way for the creation of more flexible and permeable landscapes where students optimise successfully all their potential. Invariably this leads us to discuss the concept of resilience. Even if the concept was widely developed, there is no firm background for professionals to come into practice the concept in contexts of crisis, at least in the under-developed nations (Suneeth et al., 2021). Zopiatis et al. (2021) revised the literature based on 300 already published works. They pinpoint the agenda of tourism education in five clear pillars: the capacity of forecasting, post COVID-19 curriculum, experiential learning, digital customer relations, and resilience-based practices. To put this simply, the future of tourism education rests on the interplay of service sectors, customers, and stakeholders. Tourist higher education should be tailored not only to the new times but also to the emotionality which helps to foster the most resilient attitudes before the crisis. In the next section, we describe the steps followed by PANCOE during the pandemic. Other interesting point of convergence offers the tourism epistemology; to be more exact, how can we investigate tourism in a world without tourists?

Adam Franklin coined the term tourist-centricity to exhibit the obsession of tourism epistemologists to take the tourist as the only valid source of information and knowledge. In tourism education the curricula has followed a strong imprint based on tourist-centricity which is at least dominated by marketing and management. In addition, some critical voices have alerted that tourism education has a great influence of the economic-based theory. This dominant paradigm –which monopolises an interesting number of publications and leading journals-, stipulates the training of future workforce as the main criterion towards a global curriculum (Airey et al., 2015). The COVID-19 not only engendered an unparalleled state of crisis in the tourism industry but also interrogated the current curriculum and the obsession for the economic-based paradigm (Belhassen & Caton, 2011; Korstanje & George, 2022). This moot point will be dissected and detailed with accuracy in the next section.

6.3 Challenges and Opportunities of Joy Labs and PANCOE

The methodological limitations of tourist-centricity exposed by Franklin have been manifested in two important directions. On one hand, sometimes interviewed tourists are unfamiliar with their real inner world (or emotions) whilst in other cases, they lie simply to protect their interests. What would answer a gangster if we ask: what is your profession? He would say for sure, I am a respected businessman, but indeed is he a businessman? The same applies to tourism research. Sometimes, researchers elaborate large samples applicable to tourists at bus stations or airports. These techniques are certainly based on quantitative measures but in no way do they denote scientific correlation. On another hand, how should be drawn the curriculum of a post-COVID-19 world? this is the reason behind in this second part, we lay the foundations to understand the updating of tourism curricula to educate not only tourism staff but also future scientists. Needless to say, the industry is going through serious global risks (i.e. the COVID-19 pandemic) which compromise directly its not so-long-distant future. In the next decades, professionals and practitioners should be re-trained according to the new ties in a post-COVID-19 context where adaptation plays a key role. Originally aimed at standardising students` performance through the improvement of cognitive attention, PANCOE showed outstanding results, which were already described in the book. Centred on the importance to improve students` learning skills, PANCOE combines a set of digital platforms with gamification theory. As a part of the Joy Labs, depending on the University of Palermo, Argentina, PANCOE, which started in 2012, proffers a new pedagogical model to be applied during the pandemic. The pre-pandemic days go from 2012 to 2018, and the results were shared in the previous book chapters. From 2019-to 2021, we incurred new applications and transformation of the method adjusted to the COVID-19 pandemic. Exciting and promising, the shared results cannot be extrapolated to other universes due to the sample is not statistically representative. Regardless of this point, PANCOE is situated as a finalist for the event *Reimagine Education Awards 2016* held by Wharton College, US, a great achievement that ultimately probes the efficiency of the method. Once again, PANCOE is not a method but a combination of techniques oriented to stimulate joy and wellbeing in participants.

It was not a surprise for us to see how the protocols and procedures at virtual classrooms were different in the new normal in comparison with the pre-COVID-19 days. In sharp opposition to the pre-pandemic context, education has been mainly characterised by higher levels of fear, anxiety and mistrust in which case the students´ performance and motivation have been seriously affected. As a method oriented to promote pleasure through gastronomy and heritage consumption, PANCOE should be adjusted to what experts named gamification during the COVID-19. Students were encouraged to learn whilst playing with peers in a net of new social rules and norms. In the new normal, the main goal was centred on giving participants new experiences that helped them to bolster a fluid dialogue with their peers. The sense of integration was vital to understanding such an experience.

Methodologically speaking, the sample was shaped by 50 students only who coursed 9 months of the elective year (ranging from March to November 2020). The age cohort oscillated from 18 to 13 years old. PASCOE has taken place all Mondays for two hours long whereas the rest of the week was limited to classic e-learning courses. At the preliminary facet, participants have been invited to bake pieces of bread and cook culinary dishes – needless to say that nothing of this was possible with the psychical distancing and the lockdown effective in daily life-. The activities have been expanded to acts of personal expressions such as reading a book, writing or reading poems, playing with any type of musical instrument, singing or even dancing. It is safe to say that arts have taken the lead at this stage occupying a central position in the students´ activities. At the same time, students have been asked to hang a poster on the doors of their rooms saying "*Do not disturb!. Students in class!*". In doing so, participants separated mentally from the life and activities in the rest of the home, prioritizing their attention on the task they have to accomplish. Following the same reasoning, students were invited to create a Twitter account whilst adding new followers. Over the successive weeks, they interacted directly with peers and other students who commented on their artworks. This activity facilitates them to create a dense network oriented to potentiate their motivation and engagement. PANCOE's results have been shared on Instagram and Facebook with other students and professors. PANCOE was aimed at potentiating students´ skills within the ALTAX method contributing to the creation of an individual space of self-enhancement and creativity. ALTAX method helps integrating the cognitive function through neurosciences with firm application in pedagogy and education. Each artwork has been visualised and disseminated using mass media and digital technologies. ALTAX is a method created by the University of Palermo Staff to embrace smart education. This smart education contains pleasurable experiences stimulated by endorphins which make significant dents in the circuits of the brain. Each pleasure experience not only is rememorised according to current behaviour contemplating alternative and more creative (problem solving) course of action, but also opening a fluid dialogue with the "Other". The pleasurable experience seems to be disposed through the articulation of countless practical activities such as cooking, painting, drawing, signing or writing poems or even tasting culinary dishes in view of a creative heritage. Similar to the logic of creative tourism originally described by Greg Richards (Duxbury & Richards, 2019; Richards, 2009, 2020; Richards & Raymond, 2000) ALTAX engages higher levels of solidarity amongst players. In this way, PANCOE ensured that the student's attendance (connectivity) was more than 99% with almost less than 2 absences in 9 months. Here two assumptions should be made. On one hand, it is noteworthy that students were not evaluated using the same parameters in the pre-pandemic days, so any comparison with other elective years seems to be futile. Nevertheless, PANCOE evinced higher levels of motivation and engagement which evinces the higher levels of connectivity at the virtual classrooms. As stated, there is no reliable information regarding exams and degrees during the 2020 elective year to share with our readers, but we obtained interesting information for the 2021 elective year, the moment when the restrictive measures have been removed by the government.

At a closer look, for 2015 –for further details see Table 6.1, 82% of students who have participated in PANCOE (group A) approved the elective year, which is formed by 10 annual courses whilst contrasting notably to 68% of students (in group B) who approved the year. For 2017, group A passed with almost 80% of students and B group with 65%. In 2018, group A passed successfully the year in almost 83% in comparison with group B (69%). For 2021, the year just after the pandemic, both groups have notably decreased in their approval ratio, but –this is important- PANCOE takes the lead with 75% with 67% for Group B. The same can be noted in Fig. 6.1.

Let's remind readers that each tier represents an elective year, tier 1 is for 2015, tier 2 is for 2018, two years earlier the COVID-19 pandemic. Tier 3 which notably decreases in performance is for the 2021 year, a period fraught with restrictive measures and online education. Despite the performance has been reduced, those students taking part in PANCOE have better grades than other groups (for further details see picture 1). All the discussed evidence suggests PANCOE has firm advances in enhancing the graduate students` academic performance laying the foundations towards a new understanding of tourism education.

Table 6.1 Performance of students in years 2015, 2018, 2021

Year	PANCOE - A	Group B
2015	82%	68%
2017	80%	65%
2018	83%	69%
2021	75%	67%

Source, self elaboration, PANCOE, 2012-2021

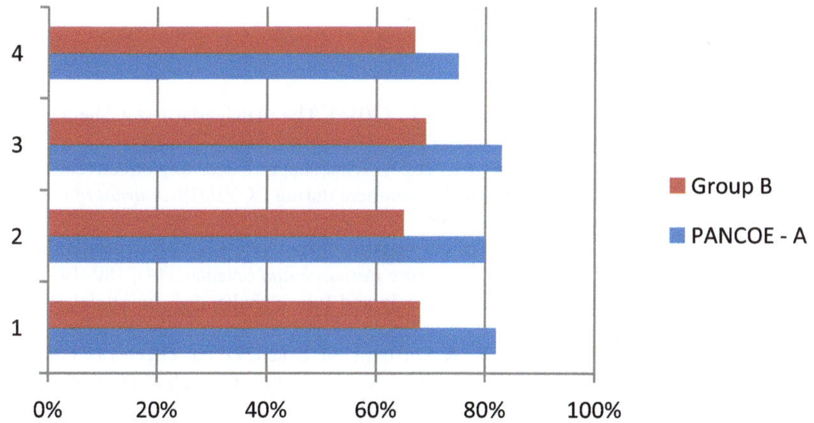

Fig. 6.1 Comparison Academic performance, Group A/B. (Source: self elaboration, PANCOE Method)

6.4 Conclusion

The virus known as SARSCOv2 (COVID-19) has unfortunately wreaked havoc not only in the tourism industry and global trade because of the restrictive measures adopted to deter the contamination but also in tourism higher education. Paradoxically, those steps and health protocols disposed by the government to stop the pandemic opened the doors to new opportunities for digital technologies and e-learning. In recent months, some experts have lamented on the psychological problems and social maladies in students who were forced to stay at home for a long period. The psychological distress caused by strictest and longest lockdowns invariably leads students to downgrade their scoring whereas increasing even the drop-out rates. In the present chapter, we introduce readers to a new pedagogic method baptised as PANCOE which is part of an experiment based on Joy Labs, University of Palermo, Argentina. PANCOE is an experiment that combines the latest advances in neurosciences with practical applications to the stimulation of pleasure and joy as positive emotions that allow the liberation of endorphins. As explained, far from what specialised literature suggests, we hold that pleasure and joy permit faster optimization of resources to improve the learning skills (even during the COVID-19 pandemic). Having said this, Joy Labs look to standardise the learning process respecting each student's cultural background. Our thesis –tested in this book- holds that those students who have taken an active part in PANCOE have not only experienced a clear reduction in the dropout rates but also have a further ration of approval in their elective years in comparison with other students who had not been involved in the experiment. This behove us to think that PANCOE's outcomes are not only outstanding but promising to be developed in future approaches and contexts mainly marked by psychologically or emotionally deprived students, in which case it includes zones of war (like Ukraine), or areas of extreme poverty and deprivation.

References

Airey, D., Tribe, J., Benckendorff, P., & Xiao, H. (2015). The managerial gaze: The long tail of tourism education and research. *Journal of Travel Research, 54*(2), 139–151.

Amin, I., Yousaf, A., Walia, S., & Bashir, M. (2021). What shapes E-learning effectiveness among tourism education students? An empirical assessment during COVID19. *Journal of hospitality, leisure, Sport & Tourism Education, 100337.*

Bakar, N. A., & Rosbi, S. (2020). Effect of coronavirus disease (COVID-19) to tourism industry. *International Journal of Advanced Engineering Research and Science, 7*(4), 189–193.

Banerjee, A. (2021). COVID-19: Fractured society and future challenges. *Industrial Psychiatry Journal, 30*(Suppl 1), S3.

Barbosa, R. B., Costa, J. H., Handayani, B., & Korstanje, M. E. (2021). The effects of COVID-19 in the tourist society: An anthropological insight of the trivialisation of death and life. *International Journal of Tourism Anthropology, 8*(2), 179–192.

Belhassen, Y., & Caton, K. (2011). On the need for critical pedagogy in tourism education. *Tourism Management, 32*(6), 1389–1396.

Buckley, R. (2021). Pandemic travel restrictions provide a test of net ecological effects of ecotourism and new research opportunities. *Journal of Travel Research, 60*(7), 1612–1614.

Chathoth, P. K., & Sharma, A. (2007). Core curricular issues in hospitality and tourism education-present structure and future directions. *Journal of Hospitality & Tourism Education, 19*(1), 10–19.

Cheng, M. T., & Agyeiwaah, E. (2021). Exploring Chinese students' issues and concerns of studying abroad amid COVID-19 pandemic: An actor-network perspective. *Journal of Hospitality, Leisure, Sport & Tourism Education*, 100349.

Collins-Kreiner, N., & Ram, Y. (2020). National tourism strategies during the Covid-19 pandemic. *Annals of Tourism Research*. https://doi.org/10.1016/j.annals.2020.103076

Duxbury, N., & Richards, G. (Eds.). (2019). *A research agenda for creative tourism*. Cheltenham.

Edelheim, J. (2020). How should tourism education values be transformed after 2020? *Tourism Geographies, 22*(3), 547–554.

Forristal, L. (2016). Using you tube videos of anthropology of tourism Pioner Valene Smith's work and philosophy to balance the tourism curriculum. In D. Prebezac, C. Schott, & P. Sheldon (Eds.), *The tourism education future initiatives: Activating change in tourism education* (pp. 130–152). Routledge.

Gössling, S., Scott, D., & Hall, C. M. (2020). Pandemics, tourism and global change: A rapid assessment of COVID-19. *Journal of Sustainable Tourism, 29*(1), 1–20.

Haryanto, T. (2020). COVID-19 pandemic and international tourism demand. *JDE (Journal of Developing Economies), 5*(1), 1–5.

Higgins-Desbiolles, F. (2020). Socialising tourism for social and ecological justice after COVID-19. *Tourism Geographies, 22*(3), 610–623.

Higgins-Desbiolles, F. (2020b). The "war over tourism": Challenges to sustainable tourism in the tourism academy after COVID-19. *Journal of Sustainable Tourism, 29*(4), 551–569.

Higgins-Desbiolles, F., & Bigby, B. C. (2022). A local turn in tourism studies. *Annals of Tourism Research, 92*(C), 1–10.

Higgins-Desbiolles, F., Bigby, B. C., & Doering, A. (2021). Socialising tourism after COVID-19: Reclaiming tourism as a social force? *Journal of Tourism Futures*. https://doi.org/10.1108/JTF-03-2021-0058

Hussain, A., & Fusté-Forné, F. (2021). Post-pandemic recovery: A case of domestic tourism in Akaroa (South Island, New Zealand). *WORLD, 2*(1), 127–138.

Islam, M., Jannat, A., Al Rafi, D. A., & Aruga, K. (2020). Potential economic impacts of the COVID-19 pandemic on south Asian economies: A review. *WORLD, 1*(3), 283–299.

Jafari, J., & Ritchie, J. B. (1981). Toward a framework for tourism education: Problems and prospects. *Annals of Tourism Research, 8*(1), 13–34.

Joshi, V. A., & Gupta, I. (2021). Assessing the impact of the Covid-19 pandemic on hospitality and tourism education in India and preparing for the new normal. *Worldwide Hospitality and Tourism Themes, 13*(5), 622–635.

Kallou, S., & Kikilia, A. (2021). A transformative educational framework in tourism higher education through digital technologies during the COVID-19 pandemic. *Advances in Mobile Learning Educational Research, 1*(1), 37–47.

Koh, E. (2020). The end of over-tourism? Opportunities in a post-Covid-19 world. *International Journal of Tourism Cities, 6*(4), 1015–1023.

Korstanje, M. E. (2020). Passage from the tourist gaze to the wicked gaze: A case study on COVID-19 with special reference to Argentina. In B. George & Q. Mahar (Eds.), *International case studies in the Management of Disasters (tourism security-safety and post conflict destinations)* (pp. 197–211). Emerald Publishing Limited.

Korstanje, M. E. (2021). Tourism disaster preparedness post-COVID-19 pandemic: The example of Argentina. In V. G. Gowreesunkar, S. W. Maingi, H. Roy, & R. Micera (Eds.), *Tourism destination Management in a Post-Pandemic Context (tourism security-safety and post conflict destinations)* (pp. 69–83). Emerald Publishing Limited.

Korstanje, M. E., & George, B. (2021a). COVID-19, terrorism, and conspiracy plot: Everyone is guilty until proven innocent. In *Mobility and globalization in the aftermath of COVID-19* (pp. 135–151). Palgrave Macmillan.

Korstanje, M. E., & George, B. P. (2021b). COVID-19 and the end of hospitality: At least as we know it in the west. In *Socio-economic effects and recovery efforts for the rental industry: Post-COVID-19 strategies* (pp. 148–164). IGI Global.

Korstanje, M. E., & George, B. (2021c). *Mobility and globalization in the aftermath of COVID-19: Emerging new geographies in a locked world*. Palgrave Macmillan.

Korstanje, M. E., & George, B. (2022). *The nature and future of tourism: A post-COVID-19 context*. Apple Academic Press.

Li, F., & Zhang, L. (2022). The influence of COVID-19 on vocational hospitality and tourism students' career choices. *Journal of China Tourism Research*, 1–18.

Liburd, J., Hjalager, A. M., & Christensen, I. (2016). Valuing Tourism education 2.0. In D. Prebezac, C. Schott, & P. Sheldon (Eds.), *The Tourism Education future Initiatives: activating change in tourism education* (pp. 107–130). Routledge.

Maital, S., & Barzani, E. (2020). The global economic impact of COVID-19: A summary of research. *Samuel Neaman Institute for National Policy Research, 2020*, 1–12.

Mao, Y., He, J., Morrison, A. M., & Andres Coca-Stefaniak, J. (2021). Effects of tourism CSR on employee psychological capital in the COVID-19 crisis: From the perspective of conservation of resources theory. *Current Issues in Tourism, 24*(19), 2716–2734.

Mohanty, P., Hassan, A., & Ekis, E. (2020). Augmented reality for relaunching tourism post-COVID-19: Socially distant, virtually connected. *Worldwide Hospitality and Tourism Themes, 12*(6), 753–760.

Mostafanezhad, M., Cheer, J. M., & Sin, H. L. (2020). Geopolitical anxieties of tourism:(Im) mobilities of the COVID-19 pandemic. *Dialogues in Human Geography, 10*(2), 182–186.

Nair, B. B. (2021). Endorsing gamification pedagogy as a helpful strategy to offset the COVID-19 induced disruptions in tourism education. *Journal of hospitality, leisure, Sport & Tourism Education, 100362*.

Norrild, J. A., & Korstanje, M. E. (2021). Maids and housekeepers at luxury hotels: Life stories in hotels of Buenos Aires, Argentina. *International Journal of Tourism Anthropology, 8*(2), 129–147.

Pappas, N. (2021). COVID19: Holiday intentions during a pandemic. *Tourism Management, 84*, 104287.

Prayag, G. (2020). Time for reset? COVID-19 and tourism resilience. *Tourism Review International, 24*(2-3), 179–184.

Qiu, H., Li, Q., & Li, C. (2021). How technology facilitates tourism education in COVID-19: Case study of nankai university. *Journal of Hospitality, Leisure, Sport & Tourism Education, 29*, 100288.

Rastegar, R., Higgins-Desbiolles, F., & Ruhanen, L. (2021). COVID-19 and a justice framework to guide tourism recovery. *Annals of Tourism Research, 91*, 103161.

Ratten, V., & Jones, P. (2021). Covid-19 and entrepreneurship education: Implications for advancing research and practice. *The International Journal of Management Education, 19*(1), 100432.

Reichenberger, I., & Raymond, E. M. (2021). Students' perceptions of tertiary tourism education and careers during the Covid-19 pandemic. *Journal of Teaching in Travel & Tourism, 21*(4), 380–402.

Richards, G. (2009). Creative tourism and local development. In R. Wurzburger, A. Pattakos, & S. Pratt (Eds.), *Creative tourism: A global conversation* (pp. 78–90). Sunstone Press.

Richards, G. (2020). Designing creative places: The role of creative tourism. *Annals of Tourism Research, 85*, 102922.

Richards, G., & Raymond, C. (2000). Creative tourism. *ATLAS news, 23*(8), 16–20.

Rogerson, C. M., & Rogerson, J. M. (2021a). Looking to the past: The geography of tourism in South Africa during the pre-COVID-19 era. In C. Rogerson & J. Rogerson (Eds.), *Urban tourism in the global south* (pp. 39–75). Springer.

Rogerson, C. M., & Rogerson, J. M. (2021b). COVID-19 and changing tourism demand: Research review and policy implications for South Africa. *African Journal of Hospitality, Tourism and Leisure, 10*(1), 1–21.

Rogerson, C. M., & Rogerson, J. M. (2021c). African tourism in uncertain times: COVID-19 research progress. *Geo Journal of Tourism and Geosites, 38*(4), 1026–1032.

Rogerson, J. M., Lekgau, R. J., Mashapa, M. M., & Rogerson, C. M. (2021). Covid-19 and local business responses: Evidence from South Africa's most tourism-dependent locality. *African Journal of Hospitality, Tourism and Leisure, 10*(1), 388–405.

Romagosa, F. (2020). The COVID-19 crisis: Opportunities for sustainable and proximity tourism. *Tourism Geographies, 22*(3), 690–694.

Ruhanen, L. (2006). Bridging the divide between theory and practice: Experiential learning approaches for tourism and hospitality management education. *Journal of Teaching in Travel & Tourism, 5*(4), 33–51.

Seo, S., & Kim, H. J. (2021). How COVID-19 influences hospitality and tourism education: Challenges, opportunities, and new directions. *Journal of Hospitality & Tourism Education, 33*(3), 147–147.

Sharma, A., & Nicolau, J. L. (2020). An open market valuation of the effects of COVID-19 on the travel and tourism industry. *Annals of Tourism Research, 83*. https://doi.org/10.1016/j.annals.2020.102990

Sigala, M. (2020). Tourism and COVID-19: Impacts and implications for advancing and resetting industry and research. *Journal of Business Research, 117*, 312–321.

Suneeth, B. G., Kashyap, S., Reddy, G. M., & Kaushal, V. (2021). Resilience adaptations in tourism education for the post-COVID-19 era–a study of India. In V. Gowreesunkar, S. Mingi, H. Roy, & R. Micera (Eds.), *Tourism destination Management in a Post-Pandemic Context* (pp. 291–302). Bingley.

Tiwari, P., Séraphin, H., & Chowdhary, N. R. (2021). Impacts of COVID-19 on tourism education: Analysis and perspectives. *Journal of Teaching in Travel & Tourism, 21*(4), 313–338.

Tzanelli, R. (2021). *Cultural immobilities and the Virocene: Mutating the crisis*. Edward Elgar.

Tzanelli, R. (2022). *Cultural immobilities and the virocene*. Edward Elgar.

Wassler, P., & Fan, D. X. (2021). A tale of four futures: Tourism academia and COVID-19. *Tourism Management Perspectives, 38*, 100818.

Wassler, P., & Talarico, C. (2021). Sociocultural impacts of COVID-19: A social representations perspective. *Tourism Management Perspectives, 38*, 100813.

Wen, J., Aston, J., Liu, X., & Ying, T. (2020). Effects of misleading media coverage on public health crisis: A case of the 2019 novel coronavirus outbreak in China. *Anatolia, 31*(2), 331–336.

Wen, J., Wang, W., Kozak, M., Liu, X., & Hou, H. (2021). Many brains are better than one: The importance of interdisciplinary studies on COVID-19 in and beyond tourism. *Tourism Recreation Research, 46*(2), 310–313.

Ye, H., & Law, R. (2021a). Impact of COVID-19 on hospitality and tourism education: A case study of Hong Kong. *Journal of Teaching in Travel & Tourism*, 1–9.

Ye, H., & Law, R. (2021b). Impact of COVID-19 on hospitality and tourism education: A case study of Hong Kong. *Journal of Teaching in Travel & Tourism, 21*(4), 428–436.

Zhang, H., Song, H., Wen, L., & Liu, C. (2021). Forecasting tourism recovery amid COVID-19. *Annals of Tourism Research, 87*, 103149.

Zhong, Y., Busser, J., Shapoval, V., & Murphy, K. (2021). Hospitality and tourism student engagement and hope during the COVID-19 pandemic. *Journal of Hospitality & Tourism Education*, 1–13.

Zopiatis, A., Pericleous, K., & Theofanous, Y. (2021). COVID-19 and hospitality and tourism research: An integrative review. *Journal of Hospitality and Tourism Management, 48*, 275–279.

Conclusion: Main Guidelines for Tourism Education

In a book entitled *global Tourism education* Cathy Hsu who is an authoritative voice in the fields of higher education, compiles a set of different works discussing the present and future of tourism education in the world. Although this book was originally written and published in 2005, which needs further updating because of recent events, some reflections are important to share. Hsu (2005) argues that the growth of the industry has generated an expansion of the higher education demand which remains unfilled by the private sector. This invariably created a lack of coordination between the public and private spheres. In the same line of inquiry, Dean Maclaurin (2005) calls attention to the Canadian study case. Per his outcomes, formal tourism education programs have more than 60 years in Canada but they were changed according to the needs of the industry. The government historically passed laws and legislations to support the industry but without any practical result. This happens simply because the public response has lagged behind the innovation and changes the private sector needs. The vulnerability of the industry to the cyclical financial downturn has been discussed by MacLaurin reminding us that the idea of creating national manpower for the sector generated a gap between labour demand and supply in the hospitality industry. In the same direction, Tom Baum (2012) analyses the case of the UK contemplating the great diversity in tourism and hospitality programs. This strength allowed not only the private sector to respond to the daily demands of the sector but also attracted foreign students. However, this diversity has opened a hot debate in academic circles just after the 2000s. The diversity given by the UK fails to train students in a standardised way or at least at international levels to meet the necessities of global destinations. Arie Reichel (2005) places the case of Israel in the foreground. The nature of higher education in Israel rests on the geopolitical tensions of the region. This led educators not only to tailor the programs to the international demand but also to introduce liberal policies connecting both spheres, private and public. The privatization of higher education has been accompanied by a technological revolution that alternates traditional with innovative learning. Phillip Pearce (2006) concludes –in the case of Australia- that the

A. Zuccoli, M. E. Korstanje, *The Role of Pleasure to Improve Tourism Education*, https://doi.org/10.1007/978-3-031-21580-3

problems of tourism higher education found in other countries are far from being gravitating in Australia. Tourism education situates for many young students a criterion of status and privilege oriented to a better position in society. Universities play an operational role –not managerial- encouraging students in applied research to meet the higher standards of the commonwealth. This particularly has attracted not only students coming from other English-speaking countries but also Asia. Education evolves according to the different stages of tourism production and consumption. It is commonly argued that the evolution of tourism has escaped the controls and borders of the classic nation-state, but what is more important, travel behaviour has notably changed over time.

As tourism-related scholars, many of us are accustomed to thinking of the industry as a fixed and static set of companies, stakeholders and consumers immutable in the threshold of time. Far from this, the tourism industry has mutated according to the different times and socio-economic dynamics from its outset. The end of WWII marked the rise of new mass tourism which accelerated not only the dynamics of globalization but also changed substantially our travel behaviour (MacCannell, 2013; Marson, 2011; Zuelow, 2015). Years of the Grand Tour or the holidays only limited by a ruling elite were over just after the 50s decade. As Eric Zuelow (2015) puts it, modern tourism is based on the condition of mass consumption which is mainly marked by the democratization of leisure practices. Although culturally rooted in the rite of Grand-tour, tourism has historically expanded because of the introduction of technological innovation in the transport system. The military dominance of Europe, determined by the colonial period, ensured a rapid connection with every corner of the globe. In this vein, John Urry (2002) coins the term "tourist gaze" to explain the fascination for gazing cultivating in western democracies. Per Urry, the tourist gaze denotes the possibility to consume "landscapes, cultures and peoples" according to a much deeper cultural matrix that dictates what can be gazed at or not. However, economies –as well as good transactions- mutate towards new forms and shapes. For Urry, tourism gradually tends to take abstract forms linked to the signs and not the materiality of the production. To put this straight, consumers (tourists) do not look to buy tangible goods but also abstract (solipsist) experiences based on the hegemony of the sign. The global modernity, where modern tourism inscribes, evolves towards a decentralised form of consumption proper what Urry dubbed as "aesthetic reflexibility" (Lash & Urry, 1993; Urry, 1992, 2001, 2002; Urry & Larsen, 2011). Similar observations have been made by senior sociologist George Ritzer and David Harvey who theorised from different angles on the rise of a new global ethos. To a major or less degree, tourism emulates the circulation and the symbolic spaces of capital which mediates between lay citizens and social institutions (Harvey, 2002; Ritzer & Dean, 2015). As a result of this, the success of capitalism in expanding the national borders crystallised in more dynamic forms of displacements and leisure travels. For instance, this evolution that seems to be grounded into a halt just after the turn of the century (2001) equates to a decentralised economy. This invariably suggests that travel behaviour and the means of production are inextricably intertwined. In the next line, we shall explain how tourism has evolved from 1945 to date. At the same time, the higher education curriculum was systematically adjusted to the new forms of travel behaviour. This conceptual model is divided into three distinguishable facets: (a) classic mass tourism, (b) sustainable tourism and (c) new (morbid) forms of tourism.

The first facet encompasses the inter-war period and the WWII end (1945) to 1970. This epoch is mainly marked by "enclave tourism" which means that local economies keep lower levels of participation in the capital exchange process. Tourist corporations invest money to develop closed destinations on former colonial islands or territories. These corporations which come from the same country as tourists repatriate the capital to their metropolis leaving a margin for locals. This form of mass tourism not only rests on the sun and beach product but also is designed under the post-Fordist scale or mass scale. In any way, tourists manifest their interest to be in contact with the locals and some of them spend their holidays in all-inclusive destinations (Britton, 1982, 1991). At the same time, the nation-state heads the first tourist campaign stressing domestic tourism (Pack, 2006). In consonance with this, it is important to add that holiday-makers in this facet are psychologically moved by the idea of relaxing and rest not only indifferent to the locals but also the undesired consequences of tourism on the environment. From 1971 to 1999 a new segment of tourism simply surfaces sustainable tourism. The term sustainability alludes to a certain type of empathy by the ecology, but this epoch resulted from a combination of two main problems that affected the capitalist form of production. The first factor dates back to the Arab-Israeli war and the derived Oil embargo in 1972 that affected the prices leading the western economies to a great economic crisis known as "stagflation". The western economies envisaged their mass scale of production and consumption cannot be sustained at the time simply because the basic commodities (oil) that greased the rails of the economy were not assured (Harvey, 1989). This was a foundational event that paved the way for the rise of new customs, like the need for segmentation and marketing to plan the next strategies to adopt. The mass-scale economy sets the pace for a decentralised economy where marketing campaigns were tailored to individual consumers. The rational planning hieratically organised has been replaced by bottom-up costumers' demands. Of course, the same applied very well to the tourism industry. The paradigm of sustainability, expressed in the Brundtland report, has taken the lead in a new agenda where the idea of conservationism and the resources for a common good prevailed (Telfer, 2012). This epoch is characterised by the expansion of the global (capitalist) economy conquering other Asian economies. Supported by the technological breakthrough, tourism expanded to the borders of the world connecting faraway cities in hours. This mobile society cultivated not only the interest for the "Other" (local) but also for consuming heritage and authenticity. Tourists are now motivated to get understanding experiences or are simply in quest of new cultures and customs to interrogate their lifestyle. Having said this, policymakers devote considerable efforts to placating the rivalry and problems between guests and hosts. At least until 2001 when the attacks on New York City, US hallmarked the third facet. During this epoch, new (morbid) forms of tourism emerged, such as creative tourism, war tourism, bottom tourism, slum tourism, dark or Thana-tourism or robot tourism (only to name a few). Undoubtedly, 9/11 not only shocked the US and the World but also configured a new policy in the geopolitical arena as never before. This was the first time, four civil airplanes are weaponised against the most powerful country in the world. This attack was unfortunately accompanied by a set of global risks such as natural disasters, the ecological crisis resulting from climate change, as well as the outbreaks of viruses (like COVID-19) and other maladies. It is time policymakers enthusiastically adopted the safety-security paradigm and risk perception theory in the fields of tourism research (Hall et al., 2004; Korstanje, 2018; Mansfeld & Pizam, 2006). In 2008 a second economic crisis whipped the capitalist system placing the tourism industry on the brink of collapse. In the mid of a political and economic crisis, tourism mutates towards a new form dotted with virtualisation and creativity. New morbid forms of tourism alter the ideals of "the beauty" introducing the "Other's suffering" as a new attraction for consumers (Korstanje, 2016). This is a type of tourism where the apollonian landscape cedes to the macabre. This is based on full decentralised planning adjoined to low-cost travels supported by digital technologies. It is the first time, so to speak, that people can engage in a journey without displacing it through virtual tourism (Molz, 2012). In this global scenario, consumers are interacting in a dense net where the classic forms of hospitality are shifted (Jordan et al., 2020). Nevertheless of this fact, the impossibility to prevent global risks led experts to imagine alternative strategies of adaptancy. What all these forms have in common is the idea of adapting or enhancing resiliency before the devastation at the time the disaster takes a hit. Some interesting studies have focused on the post-disaster or conflict destination as the most significant variable to take into consideration (Isaac et al., 2019; Seraphin et al., 2020). Having said this, the tourist products –for this epochal moment- rests on what Joseph Schumpeter baptised as "creative destruction", symbolising the importance to recycle the devastated areas of the economy or putting destruction as the precondition towards a new creation (Nicholas, 2003). As debated, higher tourism education is not aloof from these radical shifts and significant mutations in the industry. We have debated in the present book on the importance to update the curriculum content according to the new times in a post-COVID-19 world, as well as the methodological limitations of education today. It is not otiose to say that we strongly believe that COVID-19 gives us a great opportunity to overcome the barriers to tourism education towards a more integrated and negotiated curricula. Through the articulation of countless combined techniques, we amply showed how pleasure can stimulate not only academic performance, undermining students-lecturers' conflicts but also reducing the dropout rates. Having said this, PANCOE inserts within the constellations of wellness tourism toying with the belief that pleasure and joy play a leading role in the construction of a sustainable education that ultimately helps to sanitise the tourism industry. Those students who have taken part in PANCOE obtained better grades than other cohorts and groups. In turn, PANCOE was originally designed to reduce the dropout rates in foreign students or students coming from neighbouring countries. The results were exhilarating and promising, those students taking part in PANCOE experienced lower dropout rates than other samples. This suggests that pleasure occupies a central position improving not only students' academic performance but also cementing engagement. As an innovative experience, PANCOE instrumentalised successfully the students' reserve of endorphins through the stimulation of pleasure experiences such as cooking, tasting dishes, watching landscapes, baking pieces of bread, and confronting negative emotions. PANCOE directly activates to achieve a durable state of joy and well-being enhancing cognitive attention. The goals of this experiment were twofold. Primarily it was oriented to reduce the alarming dropout rates in the university, and secondly, it potentiated academic performance in psychologically-deprived students. The experiment was repeated in 2019/2020, a period mainly marked by social distance, the lockdown and the COVID-19 pandemic. Although chapters which form this book were not published in other sources, and they were exclusively written for Springer Nature, no less true is that they revolve around the same common thread of argumentation. Pleasure and joy are key factors that potentiate academic performance through the stimulation of engagement in students, not only reducing professor – students' conflict but also innovating the constellations of tourism higher education. In the present book, we have discussed the role of pleasure in potentiating higher education. The first chapter introduces readers to the problems of Western education and some vocational issues. The Cartesian dualism, which mainly marked higher education in the West, subordinates subjects' emotions (like pleasure) to a much deeper cognitive rationality. PANCOE -and of course this book- goes in an opposite direction. Pleasure has been traditionally questioned by Stoics and medieval philosophy leading the concept to a marginal position. However, and this is the point in the second chapter, problems in higher education include violence, high dropout rates, and lack of engagement and motivation in students (and so forth). The chapter certainly focuses not only on the nature and rise of modern universities but also on the conceptualization of what is a wider society. The third chapter is on the methodological limitations of tourism education. Part of these problems are similar to what we have mentioned, but to this one must add a dissociation between theory and practice. From its outset, higher tourism education emphasised the importance to provide students with abilities and skills to resolve the problems of the industry. To some extent, the current pressures on professors for gaining further positions in the metrics have invariably led them to abandon their classrooms. The turn of the twentieth century witnessed the proliferation and multiplication of countless risks which include terrorism, climate change, ecological crisis, and the recent COVID-19 pandemic. Some voices question on the importance to update the curricula given the new times. With the basis in Argentina as the main study case, the fourth chapter interrogates the problems of curricula to incorporate a new resilient agenda in higher education. The fifth and sixth chapters, lastly, exhibit a detailed review of PANCOE method, its methodological limitations, weaknesses and strengths. PANCOE was originally designed to reduce the higher dropout rates whilst homogenising the curriculum to foreign students. Both chapters have commonalities and considerable differences. Whereas the fifth chapter refers to PANCOE experience earlier during the COVID-19 pandemic, the sixth signals the outcomes obtained during 2020.

As the previous snapshot is given, it is time to embrace new paradigms associated with cognitive stimulation, creativity and lateral thinking. In so doing, PANCOE offers a fertile ground for future experimentations and debates. At least, we, the authors, this book serves educators, researchers and lecturers to lay the foundations for a new understanding of tourism education in the next decades. We wish to thank Gabriel Foglia, dean of Economics and the University of Palermo, Argentina for the generosity and the given space to orchestrate the first experiments of PANCOE as well as Springer staff (above all Natalie Rieborn, Helen Van der Stelt, Anita Rachmat & Evelien Bakker) for the opportunity our results see the light of publicity.

References

Baum, T. (2012). Global tourism higher education–the British Isles experience. In C. Hsu (Ed.), *Global tourism higher education* (pp. 39–50). Routledge.
Britton, S. G. (1982). The political economy of tourism in the third world. *Annals of Tourism Research, 9*(3), 331–358.
Britton, S. (1991). Tourism, capital, and place: Towards a critical geography of tourism. *Environment and Planning D: Society and Space, 9*(4), 451–478.
Hall, C. M., Timothy, D. J., & Duval, D. T. (2004). Security and tourism: Towards a new understanding? *Journal of Travel & Tourism Marketing, 15*(2–3), 1–18.
Harvey, D. (1989). *The condition of postmodernity an enquiry into the origins of cultural change*. Willey-Blackwell.
Harvey, D. (2002). *Spaces of capital: Towards a critical geography*. Routledge.
Hsu, C. H. (2005). Preface. In C. Hsu (Ed.), *Global tourism higher education: Past, present, and future* (pp. ziii–zixv). Routledge.
Isaac, R. K., Çakmak, E., & Butler, R. (Eds.). (2019). *Tourism and hospitality in conflict-ridden destinations*. Abingdon.
Jordan, E. J., Vieira, J. C., Santos, C. M., & Huang, T. Y. (2020). Do residents differentiate between the impacts of tourism, cruise tourism, and Airbnb tourism? *Journal of Sustainable Tourism*. ahead of print, 1–19.
Korstanje, M. E. (2016). *The rise of tourism and Thana-capitalism*. Routledge.
Korstanje, M. E. (2018). *Terrorism, tourism and the end of hospitality in the 'West'*. Palgrave Macmillan.
Lash, S., & Urry, J. (1993). *Economies of signs and space* (Vol. 26). Sage.
MacCannell, D. (2013). *The tourist: A new theory of the leisure class*. University of California Press.
MacLaurin, D. (2005). Tourism education in Canada: Past, present, and future directions. In C. Hsu (Ed.), *Past, present, and future* (pp. 1–26). Routledge.
Mansfeld, Y., & Pizam, A. (Eds.). (2006). *Tourism, security and safety*. Abingdon.
Marson, D. (2011). From mass tourism to niche tourism. *Research Themes for Tourism, 1*(1), 1–11.
Molz, J. G. (2012). *Travel connections: Tourism, technology and togetherness in a mobile world*. Routledge.
Nicholas, T. (2003). Why Schumpeter was right: Innovation, market power, and creative destruction in 1920s America. *The Journal of Economic History, 63*(4), 1023–1058.
Pack, S. (2006). *Tourism and dictatorship: Europe's peaceful invasion of Franco's Spain*. Springer.
Pearce, P. L. (2006). Australian tourism education: The quest for status. In C. Hsu (Ed.), *Global tourism higher education: Past, present, and future* (pp. 251–268). Routledge.
Reichel, A. (2005). Tourism and hospitality higher education in Israel. In C. Hsu (Ed.), *Past, present, and future* (pp. 61–88). Routledge.
Ritzer, G., & Dean, P. (2015). *Globalization: A basic text*. Wiley.
Seraphin, H., Korstanje, M., & Gowreesunkar, V. (2020). Diaspora and ambidextrous management of tourism in post-colonial, post-conflict and post-disaster destinations. *Journal of Tourism and Cultural Change, 18*(2), 113–132.
Telfer, D. (2012). The Brundtland report (our common future) and tourism. In D. Telfer (Ed.), *The Routledge handbook of tourism and the environment* (pp. 235–248). Routledge.

Urry, J. (1992). The tourist gaze “revisited”. *American Behavioral Scientist, 36*(2), 172–186.
Urry, J. (2001). Globalising the tourist gaze. In B. Sutheeshna, M. Sitikantha, & B. Bivraj (Eds.), *Tourism development revisited: Concepts, issues and paradigms* (pp. 150–160). Sage.
Urry, J. (2002). *Consuming places*. Routledge.
Urry, J., & Larsen, J. (2011). *The tourist gaze 3.0*. Sage.
Zuelow, E. (2015). *A history of modern tourism*. Macmillan International Higher Education.

EPILOGUE: Something to Teach: An Auto-Ethnography

Introduction

The present book chapter synthesises the experiences and narrations collated as long as an auto-ethnography once I taught philosophy in some areas of public universities where tourism-related careers are taught. We made the final decision to place this chapter as a corollary or an epilogue simply because it shed a light of inspiration for PANCOE. Needless to say this ethnography was conducted some years earlier the application of PANCOE. Anyway, we believe that it reflects the factors that determine the students' motivation and engagement. Let us remind readers that in Argentina, like many other Latin American countries, policy-makers and scholars are worried about the ethical problems beyond the tourism industry as well as the problems in the higher education system. To the student's lack of interests for curricula one must add the higher dropout rates (Korstanje, 2010). Unless regulated, the sector generates some negative effects on the community such as inflation, unemployment, forced migration or even real-estate speculation. Students are also educated to deal with the problems of the sector but paradoxically less is known about their fears, expectations and worries generating a real professional paradox. This paradox is given by the fact that students are encouraged to be key players in the decision-making process of the sector, but this never happens in reality. Students once their degree is earned, far from situating in a managerial position, are subject to working excessive hours whilst low paid. In this vein, new policies should be issued within the higher education system. The first section deals with the importance of ethics introduced in the fields of tourism and hospitality higher education. There is a clear dissociation between theory and practices in tourism education. Often students are educated and trained in ethics but rarely are they applied once they become practitioners. What is more important some studies have recently questioned to what extent tourism and business are ethical activities. Complementarily, the second part explores what I dubbed as the professional paradox which reflects

A. Zuccoli, M. E. Korstanje, *The Role of Pleasure to Improve Tourism Education*, https://doi.org/10.1007/978-3-031-21580-3

the fears and expectances of graduate students during their final courses. Last, I give some description and important comments in my out obtained outcome whilst giving a course of deontology and ethics in tourism-related careers. Beyond the overt lack of enthusiasm found in students, there are countless problems mainly associated with the lack of self-esteem and unresolved psychological frustrations as well as the tension between students and lecturers. Put simply, some voices have catalogued the auto-ethnography as an unethical form of inquiry simply because laypeople remain incognizant they are being studied. Other scholars enthusiastically embrace autoethnography to study slippery matters otherwise would be inexpungable to the human eye. These matters include social conflict, prejudice and racism and illegal activities. In our respect, auto ethnography does not violate any ethical concern given the fact that subjects' integrity or their identities have not really revealed [for further detail see on De Volo & Schatz, 2004; Katz, 1997; Atkinson, 2007). In fact, like the above-noted title reflects, part of the current crisis in tourism education appears to be explained in that universities have nothing important to teach.

Ethics in Tourism and Hospitality Bachelors

At a closer look, the problem of ethics in tourism development was widely discussed and studied worldwide. Ethics play an important role in the configuration of more sustainable destinations (Duffy & Smith, 2004; Fennell, 2015; Holden, 2003). However, ethics are regularly introduced in the tourism education curricula (Fennell, 2015). One of the pioneering voices in discussing ethics in tourism education was Linda Enghagen who in 1990 published a more than interesting paperwork entitled Teaching ethics in hospitality and Tourism Education. Her argument holds that the act of teaching ethics takes two dimensions. One dimension refers to the urgency to impart a specific body of knowledge strictly linked to ethical analysis. The second one, rather, signals the challenge to invite students to act ethically –even when they are graduated. The success in teaching ethics should be measured not only by the goals reached by students in the course but also by their professional performance. Ethics often interrogates vital questions regarding what extent my decision is legal or mix-balanced for the system. Here a paradoxical situation surfaces, there can be no assurance that students will act ethically when they get managerial positions or an important position in a corporation. As the author goes on to write: "*the challenge of teaching ethics is to successfully move from imparting this knowledge about ethics and ethical analysis to motivate students to do ethically. To appreciate the difficulty of this undertaking, it is necessary to develop some understanding of theories of moral development*" (Enghagen, 1990: 470).

Having said this, she holds that in the long-term, educators should flesh out new techniques and methods to measure how ethically professionals behave. Sheldon and Fesenmaier (2013) acknowledge that ethics occupies a central position in career development simply because it ensures the sustainability of the industry in the long run. In an uncertain world fraught with global dangers and constant political change, policymakers should adopt a clear ethical model to pass laws and policies that help minimise the costs for society. Here we beg two questions: to what extent the quest for profitability may lead towards an ethical answer? and is tourism an ethical activity?

Peris-Ortiz et al. (2021) try to respond to the first above-mentioned point. Ethical behaviour has been widely analysed in the literature over decades but unfortunately, there are more doubts than answers. One of the limitations of applied research in ethics consists in the multifaceted aspects of stakeholders and agents who seek to maximize their gains whilst minimising costs. Even inside a corporation, the evidence suggests that upper management tends to protect its interests and priorities no matter the consequences in the company. Authors prefer to use the term Ethical business behaviour to denote the type of communal practice oriented to achieve a common good. The introduction of entrepreneurship not only helps to solve the practical problems of the community but allows convergence between gains and reciprocity, as the authors conclude. There is a critical point of entry in this discussion between self-interest and the common good. Upper management behaves selfishly simply because they have a favourable environment but things change when the resources are exhausted –or problems remain unresolved-. In consonance with this, Álamo et al. (2021) bring a critical reflection to the extent firms or corporations behave ethically whilst amassing profits. Per their viewpoint, the debate seems to be far from closed. Like humans who had the decision to do a thing good or bad, corporations have the opportunity to behave ethically or not. This means that firms not only have their freedom but also stretch out the three factors of ethics: knowledge, intention and previous awareness of the possible effects of the decision making process. If we toy with the belief that the ultimate goal of humans is happiness, the goal to reach by firms is value creation. Of course, even if freedom is essential for ethics it is not limited to. Some corporations are oath to introduce radical shifts towards entrepreneurship or a more sustainable society. Firms moving within the sphere of law looking for gains without affecting the common good can be valorised as ethical ones. One might speculate that the capacity to solve problems creatively (entrepreneurship) keeps a mix-balanced state where profits and ethics coexist. Interesting studies of entrepreneurship applied to tourism business have been published by Ateljevic & Doorne, 2000; Zhao et al. (2011), Ratten (2019), Ratten et al. (2019). All these works emphasise entrepreneurship not only as an emerging area or object of study for professional researchers but as a fertile ground towards a more sustainable society. The concept of social capital occupies a central position in the formation of entrepreneur leaders as well as their skills to face adversities. For some reason, ethics in tourism higher education advances at a snail's pace. As David Fennell (2015) puts it, ethics in tourism education is based on a level of complete disconnection revolving around a chasm between the learning process and research. This disconnection makes it very important for policymakers –and above all educators- to adopt ethics in the tourism curriculum. Per his viewpoint, the learners of today are the practitioners of tomorrow. Last but not least, beyond the importance of ethics in tourism education, the interests of educators seem to be slim. This point leads us to hold that the sustainability of tourism education rests clearly on shaky foundations (Boley, 2011; Deale et al., 2018; Tribe, 2002). In the next section, we shall discuss what we dubbed as the professional paradox, which explains the higher dropout rates in tourism and hospitality. Students are often educated to fill significant positions in the system but they are ultimately subjects to inhospitable working conditions. The idea that professionalisation in tourism is not necessary, or students are trained only to work as professionals (not scientists or researchers) is one of the topics we bring into the foreground. The paradox lies in the fact the same motivations students have in their pre-graduate courses become in frustration at the threshold of their career.

The Professional Paradox: Scientists or Tour Guides?

Over the recent decades, the growth of the tourism industry, as well as the application of technologies to accelerate travel times, have created new forms of discourses, mobilities and emotions that affected directly education (Andrades & Dimanche, 2017). The multiplication of Ph. Doctors, masters and Bachelors with a focus on tourism and hospitality is an unquestionable reality. English speaking countries (like the US, the UK or Australia) has taken the lead in the formation of future tourism researchers, but other developed countries such as China, Japan, Germany, France without mentioning Spain and Portugal have achieved interesting progress in this matter (Ayoun et al., 2010; Busby, 2003; Busby & Gibson, 2010; Richards, 2006; Zhao, 1991). The philosophical dilemma of higher tourism education to train future tour guides (professionals) or scientists in the fields of tourism was widely scrutinised by Professor John Tribe. Per his standpoint, he acknowledges that the managerial gaze, which means the monopolization of management and marketing in the curricula formation, is impacting higher education globally. One of the divergent points of discussion between scientists and policymakers has been pointed out by Tribe. Whilst the former looks for describing objective facts, the latter work harder to achieve a common good. Although this point was already addressed in earlier sections, in the present one it would be debated in more detail. The multiplication of tourism research not only has been subject to metrics but has also affected academic performance. The economic downturn that whipped the US and the UK in 2008 has invariably constrained the public and private funding invested in higher education. As a result of this, further levels of accountability are requested by authorities. It is important to add that universities simply reacted by increasing the fee and tuition fee to students which led to a much deeper lack of interest for graduates for tourism careers. At the same time, external scrutiny over the teaching process as well as academic performance has taken place in recent decades, creating a climate of competition and pressure for researchers. What is equally important, all these changes adopted by authorities gravitated directly to the quality of the higher education system and the students' curricula. To put the same in bluntly, higher education enthusiastically embraced a managerial gaze as it was originally designed to measure the academic and students' performance. Based on bureaucratic methods of scrutiny, these policies harmed the quality of the learning process. Other significant accelerated by the metrics go in the direction of student satisfaction. As Tribe eloquently observes, the levels of student satisfaction have slumped down just after the introduction of metrics and managerial methods in English speaking countries. Without any doubt, this reality represents a challenge posed on the tourism academy (Airey et al., 2015). This begs the question to what extent Academy is responsible for this situation?

In different texts, Tribe alert on the problem of "tourism indiscipline" which derives from two combined factors. The lack of interests of the Academy to set a clear agenda to fame knowledge production and research, associated with the multiplication of a disordered state of knowledge mainly marked by isolated academic tribes with low levels of permeability. These tribes, far from bolstering a fluid dialogue, work on their own (closed) paradigms and sun-disciplines whilst engendering what he termed as "a state of indiscipline" (Tribe, 1997, 2010; Tribe & Liburd, 2016). Starting from the premise that tourism students have a vocational crisis, Tribe holds the thesis that universities has easily ushered and digested by managerial practices mainly based on business and profits. At a closer look, the neoliberal agenda rushes universities to be highly competitive to educate entrepreneur leaders. These market-led policies have serious effects on the curriculum simply because students are trained to serve the greater good. In consequence, students, as well as educators, come through a process of de-professionalisation stigmatised by *commoditised marketable knowledge* (Tribe, 2014). It is important to mention, all these limitations (which include students' motivation and the decline of recruitment rates-aggravate the rise of the 2008 economic crisis (Ayikoru et al., 2009). Pizam and Shani (2009) manifest some concern given the discrepancy of definition practitioners and academicians have revolved around terms such as tourism or hospitality. In more than pungent paperwork, they claim that practitioners are inclined to adopt different definitions of tourism than academicians. Following the same line of reasoning, professional managers in the sector develop a different perspective than graduate students. In a combination of interviews with managers and students, authors give some interesting topics which merit to be discussed. On one hand, managers lament that the tourism industry keeps old-fashioned practices mainly marked by the lack of technological updating. Given this fact, technology allows the resolution of a global sector characterized by more demanding and sophisticated customers. Since the quality of service is in constant evolution, technology should be oriented to maximise customers' pleasure. In this way, profitability and hospitability have not further tension through managers' perspectives. Rather, students exhibited opposite emotions and dispositions once interviewed. Students overtly claimed that the industry is far from innovative because it is often the prey of repetitive and bureaucratic procedures and protocols. The act of being open-mind to customers or hospitable seems not to be appropriate for all psychological profiles. Other students showed their discontent given what they called "emotional labour" which means the concealment of sacrificing their feelings to satisfy customers' desires. Besides, the excessive working hours that prevent an alternative lifestyle was mentioned by students in the fieldwork. Pizam & Shani conclude that students give further priorities to the working conditions as well as the emotional factors in the industry whilst managers prioritise the economic and developmental factors.

As the previous argument is given, the psychological frustration in tourism students may very well lead to lower degrees or even the decision of leaving the career. The problem lies in the fact the real reasons for internship dropout rates are very hard to precise. Educators and specialists of all pundits often manipulate statistics to validate their previous ad-hoc hypotheses whilst in others they enthusiastically embrace clichés or ideological slogans like the updating of the curriculum or the decline of public funding to explain the students' distress. Under some conditions which are very hard to precise here, experts recur to external information instead of delving into the students' inner world (Richey et al., 2019). In tourism and hospitality, this issue remains particularly unexplored or under-investigated. As Anastasios Zopiatis (2007) describes, an internship can be defined as a practice where students gain further experience in the selected career or area of expertise contributing –in some cases- to earn money or some reward. This experience helps students to apply their theoretical framework to practical problems in the subsector. To put the same in other terms, the figure of internship mediates between theory and practice. The ongoing demands in the tourism and hospitality sectors need the incorporation of new alternative practices to train students as the future workforce. To some extent, nowadays all tourism programs contemplate some internship in their content globally. Internship –at least theoretically- provides the student with further self-confidence whilst enhancing employability and new professional opportunities. At the same time, the industry recruits low-cost workers who would act as the next generation of managers optimizing the gains and training selection. Under the supervision of educational establishments, internship contributes substantially to innovative learning. However, as Ziopiadis laments, this process is fraught with problems, restrictions and limitations. Whilst hospitality and tourism professionals often claim a lack of experience, students manifest their discontent because they are derided or discriminated against by their colleagues or simply fail to find a place in the firm. For the instance, expressions of xenophobia, language barriers or even cultural incompatibilities are found by specialists as a key factor towards students' frustration (Lyons & Branston, 2006). Similar observations have been directly correlated with Asian students once inserted in Western landscapes (Carvalho, 2021; Farmaki, 2018; Ruhanen et al., 2013). Although the literature abounds, vocational problems in tourism students seem to be covered in academia (Chang & Hsu, 2010). In this vein, Te-Yi Chang & Jui-Man Hsu proffer an interesting analysis that helps to understand this much deep-seated issue. Centred on the study case of Taiwan, they hold that the growth of the tourism industry has generated internally further necessities to adapt the current workforce to the new globalised forces of the market. In so doing, governments often disposed of legal reformation to give the sector more competition and efficiency. Nonetheless, the multiplication of educational offerings not only materialises with the desired results. The efforts to improve the students' curricula coincide with insufficient industrial cooperation without mentioning discrimination. To solve the problem, as the authors suggest, the industry should acquire the latest skilled experience emanated from the academic circles, whilst academicians should be emphatic with the troubles the industry often experiences. In a nutshell, students should be measured and evaluated following new (emerging) criteria accelerating drastic changes not only in the tourism higher education system but also in the vocational and quality assurance method. For some reason, probably given by the Cartesian dualism in tourism education, emotions are not being taken seriously in consideration in these types of diagnoses. The cognitive skills, as well as the academic performance, occupy a central position in the formation of students' curricula. Nonetheless of this fact, some voices have alerted on the importance to adopt students' emotions in the analysis of the learning process. By the side, Kim and Davies (2014) report positive results when curricula take a focus centred on students' learning, which not only stimulates motivation but enhances engagement, however, some structural problems and difficulties are identified. Problems in education are mainly associated with disjoined engagement and poor academic performance due to excessive levels of uncertainty or uncertain career aspirations. The specialised literature suggests that low aspirations dominate in tourism careers and among students. They keep little interest in working in the tourism industry as a long-career alternative, in which case not only the employability but also employers' perception is harmed (Ramakrishnan & Macaveiu, 2019).

Having said this, Walker and Manyamba (2020) conducted empirical-based research to ponder the degrees of discomfort and frustration in students. Such an emotion-focused approach stresses recent developments in transformative learning to improve classroom education. Based on discomfort as the main emotional drive, authors are consistent with the idea that discriminatory practices –above all detected in international internships- wreak havoc on the students' creativity. In this token, Anna Farmaki (2018) argues convincingly that bad experiences in international internships increase the probability to abandon the career in the long run. Not surprisingly, despite the benefits of internships reported worldwide, no less true appears to be that negative internship experiences recently show how students fail to finish their careers. To cut the long story short, frustrations and fears in students –as the main object of study of experts, educators- seems to be at a standstill. An auto-ethnography conducted in 2010 reveals two important things about students' frustrations: the lack of interest in the curricula, as well as fears to earn the degree.

A Study Case-Based with an Auto-Ethnography

The present section reflects my own experience as a philosopher who has been designed to teach courses of ethics and deontology (two classic philosophical traditions) in the tourism and hospitality fields. In 2010, I was contracted to give philosophy during a year for a well-famous private institute strictly dedicated to tourism-related careers. Originally, I was appointed to give these lectures to two courses; so to speak, each one ranging from 15 to 25 students respectively. Whilst drinking a coffee –the first day- I bolstered a fluid dialogue with my colleagues who harped on in the lack of interest of students for philosophy and social sciences. At the same time, a whole portion of them not only showed low degrees but also lower levels of participation in the classroom. At a first glimpse, I have no idea on the real causes of this problem but supposed lecturers (earlier than me) evinced some types of personal frustration directed against their students. I was afraid not only to cause a bad impression in my first appearance but also to be ranked as a bad lecturer. Having said this, in the next lines, I shall describe with some detail my auto-ethnography as a lecturer of philosophy in tourism careers of a private institute geographically located in Buenos Aires, Argentina.

Named as Ethics and Deontology of Tourism, my class started at 8.00 AM 2 days a week. The syllabus content was drawn according to specific and successfully study cases reported in the constellations of tourism and hospitality. I also combined stoic philosophy as well as epicurean tradition –with a focus on the role of joy and pleasure- with reading analysis. When I opened the doors, I found that students showed a clear antipathy for my class. I have been warned simply because after my formal presentation they did not show any type of insubordination (as my colleagues insisted) for me. I prepared the class to discuss the intersection between epicurean philosophy and the complex interplay between pleasure and morality. Even if they were certainly captivated by my speech, they had no previous education in philosophy or sociology. One of them asked me: *is this a class of tourism? I am here to be trained as a tour guide, nor an analytical philosopher? Tourism is a leisure activity oriented to travel around the world.*

Unnerved by this assertion, I realised my students have no previous familiarization with a clear definition of ethics. Just after the class ended, I understood that my collages have exaggerated their impression. Students worked hard to understand my successive speech even when they were incognizant with the world of Plato, Epicurus or even Aristotle. In the next encounter, students demonstrated interest to take part in the discussions, revolving around the role played by pleasure in Epicurean philosophy, as well as the risks or dangers the excess of pleasure evinces. For some reason, these motivated students engaged heavily with the classrooms whilst we expanded the recommending literature and readings. The evidence before me struck me that my colleagues have nothing to teach. Beyond the so-called lack of interest, I found that they were afraid to graduate because of the poor labour conditions of the sub-service sector. Most of them manifested clear distress due to the hostile labour conditions as well as the competitive factors expressed in the liberal market. For that reason, two questions immediately emerged: why do lecturers gossip about their students? And why are students frightened to be graduated?

The answer lies doubtless in what I dubbed as "the professional paradox". For the sake of clarity, any act of separation causes extreme panic given the fact our phenomenological world splits in two, here-and-now, and there-and-tomorrow. As a rite of passage, any graduation means a symbolic separation to the student to be re-introduced in a new (unknown) professional status. Serving as professionals, students are subject not only to a new landscape but also to new conditions of competition. Like a journey, graduation entails the rise of ambivalent emotions. On one hand, students express their fear to be recruited as front-desk staff dealing with clients and customers. On another, they want to live up to the trust of their relatives who supported them during their careers. The final exams aggravate these fears and some students leave their careers. Paradoxically, the rite of graduation starts a new process –students often resist- but at the same time, they need to do it to pass the next stage of their formation. These ambivalent feelings cause higher levels of antipathy and fear which should be rechanneled by the lecturer. Last but not least, in one of my conversations with these students, they said overtly on the fact they were originally moved to study tourism –with a focus on management issues- but not for working as a tour guide. It was unfortunate that in Argentina only two establishments (very expensive by the way) offered some access to Bachelors without any intermediate degree of Tour-guide. Although my students do not like to serve as guides, they were obliged to do so to cost their studies and the rest of their careers. As a result of this, most of them would never reach their tourism BA as planned. This professional paradox seduces students to be part of an illusion –above all they enrol at the university to act as policy-makers or top-managers in tourist organizations- but they are pressed to re-adapt the aspirations coping with intermediate degrees to absorb the tuition fees. Behind their so-called lack of motivation, students are ultimately subject to countless barriers and problems once they enter the labour marketplace. Last but not least, the present section exhibits the vocational dichotomies beyond the lack of motivation in students, problems unless otherwise resolved will affect seriously tourism higher education quality. At the same time, it brings some reflection to unpack the unspoken hostility between lecturers and students.

Conclusion

Having discussed the crisis of tourism education in earlier chapters, as well as the Cartesian dualism without mentioning the economic-based theory which prioritises profits over other criterion, the present book chapter was aimed at exploring the micro-sociological factor beyond students' psychological frustration. It is well known that dropout rates are increasing in many Western universities whilst the students' performance has in decline. The literature gives some macro-structural insights associated with the growth of tourism industry, the high demand of tourism-career to fill positions in the industry, the necessities to update the curriculum as well as the funding factor which ignites a hot debate revolving around the public vs. The private university. Beyond this dualism, this chapter discussed critically the problem of tourism education with strong focus on the students, their fears, hopes and problems during the careers. We have provided an auto-ethnography conducted during 2010 in a well-known educational establishment dedicated to tourism and hospitality. We also coin the term *professional paradox* to denote a situation where students are forced to finish their careers but this pressure puts in a dilemma; their professional future in an industry which is not prepared to receive them. Several of the reviewed papers and books in this section stress on the students' frustrations once graduated because their expectances are far from being met by managers and practitioners. The opposite is equally true, an indifferent industry which ignores students are doomed to collapse in view of the next challenges and difficult situations in the years to come.

References

Airey, D., Tribe, J., Benckendorff, P., & Xiao, H. (2015). The managerial gaze: The long tail of tourism education and research. *Journal of Travel Research, 54*(2), 139–151.

Álamo, P., Del Valle, M. D. C. E., & Lis-Gutiérrez, J. P. (2021). Firms as moral agents in the COVID-19 era: Ethical principles that shall guide the Company's relationship with its stakeholders. In M. Peris-Ortis, P. Marquez, J. Alonso Gomez, & M. Lopez Sieben (Eds.), *Progress in ethical practices of businesses* (pp. 13–44). Springer.

Andrades, L., & Dimanche, F. (2017). Destination competitiveness and tourism development in Russia: Issues and challenges. *Tourism Management, 62*, 360–376.

Ateljevic, I., & Doorne, S. (2000). 'Staying within the fence': Lifestyle entrepreneurship in tourism. *Journal of Sustainable Tourism, 8*(5), 378–392.

Atkinson, P. (2007). *Ethnography: Principles in practice*. Routledge.

Ayikoru, M., Tribe, J., & Airey, D. (2009). Reading tourism education: Neoliberalism unveiled. *Annals of Tourism Research, 36*(2), 191–221.

Ayoun, B., Johnson, M. K., Vanhyfte, M., & O'neill, M. (2010). A comparison study of US and non-US education internationalization practices of hospitality and tourism programs. *Journal of Teaching in Travel & Tourism, 10*(4), 335–361.

Boley, B. B. (2011). Sustainability in hospitality and tourism education: Towards an integrated curriculum. *Journal of Hospitality & Tourism Education, 23*(4), 22–31.

Busby, G. (2003). Tourism degree internships: A longitudinal study. *Journal of vocational education and training, 55*(3), 319–334.

Busby, G. D., & Gibson, P. (2010). Tourism and hospitality internship experiences overseas: A British perspective. *Journal of Hospitality, Leisure, Sports and Tourism Education (Pre-2012), 9*(1), 4–8.

Carvalho, I. (2021). "It's nice when they speak back to you in Chinese!" Frustration, perseverance, and linguistic accommodation in language travel. *European Journal of Tourism Research, 28*, 1–7.

Chang, T. Y., & Hsu, J. M. (2010). Development framework for tourism and hospitality in higher vocational education in Taiwan. *Journal of Hospitality, Leisure, Sports and Tourism Education (Pre-2012), 9*(1), 101–109.

De Volo, L. B., & Schatz, E. (2004). From the inside out: Ethnographic methods in political research. *PS: Political Science & Politics, 37*(2), 267–271.

Deale, C. S., Schoffstall, D. G., & Lee, S. H. (2018). Perceptions of ethical issues in hospitality and tourism research. *Journal of Hospitality & Tourism Education, 30*(4), 203–214.

Duffy, R., & Smith, M. (2004). *The ethics of tourism development*. Routledge.

Enghagen, L. K. (1990). Teaching ethics in hospitality & tourism education. *Hospitality Research Journal, 14*(2), 467–474.

Farmaki, A. (2018). Tourism and hospitality internships: A prologue to career intentions? *Journal of Hospitality, Leisure, Sport & Tourism Education, 23*, 50–58.

Fennell, D. A. (2015). Ethics in tourism. In G. Moscardo & P. Beckendorff (Eds.), *Education for sustainability in tourism* (pp. 45–57). Springer.

Holden, A. (2003). In need of new environmental ethics for tourism? *Annals of Tourism Research, 30*(1), 94–108.

Katz, J. (1997). Ethnography's warrants. *Sociological Methods & Research, 25*(4), 391–423.

Kim, A. K., & Davies, J. (2014). A teacher's perspective on student centred learning: Towards the development of best practice in an undergraduate tourism course. *Journal of Hospitality, Leisure, Sport & Tourism Education, 14*, 6–14.
Korstanje, M. (2010). Something to learn: A conflictive relationship among students and lecturers. *Educational Research and Reviews, 5*(1), 010–015.
Lyons, A., & Branston, C. (2006). Cross cultural change, adjustment and culture shock: UK to USA. *Tourism: An International Interdisciplinary Journal, 54*(4), 355–365.
Peris-Ortiz, M., Márquez, P., Gomez, J. A., & López-Sieben, M. (2021). Ethical dimension in corporate and management behaviour. In M. Peris-Ortiz, P. Márquez, J. A. Gomez, & M. López-Sieben (Eds.), *Progress in ethical practices of businesses* (pp. 1–12). Springer.
Pizam, A., & Shani, A. (2009). The nature of the hospitality industry: Present and future managers' perspectives. *Anatolia, 20*(1), 134–150.
Ramakrishnan, S., & Macaveiu, C. (2019). Understanding aspirations in tourism students. *Journal of Hospitality and Tourism Management, 39*, 40–48.
Ratten, V. (2019). Tourism entrepreneurship research: A perspective article. *Tourism Review, 75*(1), 122–125.
Ratten, V., Costa, C., & Bogers, M. (2019). Artisan, cultural and tourism entrepreneurship. *International Journal of Entrepreneurial Behavior & Research, 25*(4), 582–591.
Richards, G. (2006). Tourism education in the new Europe. In D. Hall, M. Smith, & Marciszweska (Eds.), *Tourism in the new Europe: The challenges and opportunities of EU enlargement* (pp. 52–64). CABI.
Richey, J. E., Andres-Bray, J. M. L., Mogessie, M., Scruggs, R., Andres, J. M., Star, J. R., et al. (2019). More confusion and frustration, better learning: The impact of erroneous examples. *Computers & Education, 139*, 173–190.
Ruhanen, L., Robinson, R., & Breakey, N. (2013). A foreign assignment: Internships and international students. *Journal of Hospitality and Tourism Management, 20*, 1–4.
Sheldon, P. J., & Fesenmaier, D. R. (2013). The tourism education futures initiative (TEFI): Activating change in tourism education. In I. Altejevic, N. Morgan, & A. Pritchard (Eds.), *The critical turn in tourism studies* (pp. 117–137). Routledge.
Tribe, J. (1997). The indiscipline of tourism. *Annals of Tourism Research, 24*(3), 638–657.
Tribe, J. (2002). Research trends and imperatives in tourism education. *Acta Turistica*, 61–81.
Tribe, J. (2010). Tribes, territories and networks in the tourism academy. *Annals of Tourism Research, 37*(1), 7–33.
Tribe, J. (2014). The curriculum: A philosophic practice? In D. Dredge, D. Airey, & M. Gross (Eds.), *The Routledge handbook of tourism and hospitality education* (pp. 49–61). Routledge.
Tribe, J., & Liburd, J. J. (2016). The tourism knowledge system. *Annals of Tourism Research, 57*, 44–61.
Walker, J., & Manyamba, V. N. (2020). Towards an emotion-focused, discomfort-embracing transformative tourism education. *Journal of Hospitality, Leisure, Sport & Tourism Education, 26*, 100213.
Zhao, J. L. (1991). A current look at hospitality and tourism education in China's colleges and universities. *International Journal of Hospitality Management, 10*(4), 357–367.
Zhao, W., Ritchie, J. B., & Echtner, C. M. (2011). Social capital and tourism entrepreneurship. *Annals of Tourism Research, 38*(4), 1570–1593.
Zopiatis, A. (2007). Hospitality internships in Cyprus: A genuine academic experience or a continuing frustration? *International Journal of Contemporary Hospitality Management., 19*(1), 65–77.

MIX
Papier aus verantwortungsvollen Quellen
Paper from responsible sources
FSC® C105338

If you have any concerns about our products,
you can contact us on
ProductSafety@springernature.com

In case Publisher is established outside the EU,
the EU authorized representative is:
Springer Nature Customer Service Center GmbH
Europaplatz 3, 69115 Heidelberg, Germany

Printed by Libri Plureos GmbH
in Hamburg, Germany